COMMERCIAL
TIMBERS
OF
THE WORLD

Fifth edition

Commercial Timbers of the World

Fifth edition

Douglas Patterson

Gower Technical Press

Previous editions by F.H. Titmuss published by Technical Press

Fifth edition published 1988 by
Gower Technical Press Ltd,
Gower House,
Croft Road,
Aldershot,
Hants GU11 3HR,
England

Gower Publishing Company,
Old Post Road,
Brookfield,
Vermont 05036,
USA

British Library Cataloguing in Publication Data

Patterson, Douglas
   Commercial timbers of the world—5th ed.
   1. Timber
   I. Title   II. Titmuss, F.H.
   674'.132   TA420

   ISBN 0–291–39718–2

Printed in Great Britain at the
University Press, Cambridge

Contents

Preface

Since the last edition of this book was published in 1971 F.H. Titmuss has sadly died. The publishers have decided to produce a new 5th edition; having known 'Tim' Titmuss I am proud to have been asked to write it.

There has been a gradual widening of the use of timber species and so a need has arisen for a general statement of timber processing and properties, rather than a concentration on architectural and engineering requirements only. This has led to a different presentation of the opening sections.

The identification of timbers is difficult unless the reader has a fair understanding of anatomical structure. It is not easy to provide for this within a book of this size, so no attempt has been made. Instead, a drawing of the smoothly cut end-grain of many important species is provided; the reader can compare the trimmed end-grain of an unknown or doubtful piece with the pictures and judge for himself how good a match it is. If any doubt remains then the advice of experienced people should be sought, such as the Timber Research and Development Association, Hughenden Valley, High Wycombe, Buckinghamshire or the Advisory Services of the Building Research Establishment, Garston, Watford.

Some of the timbers included in earlier editions have become more important in recent years and a few are no longer heard of. The entries have therefore been extended or omitted accordingly. The market has in general settled and few new species need to be added.

A system of notation has been introduced which highlight certain important features, such as working properties and durabilities.

Acknowledgements

Thanks are due to my son, William Ntim-Patterson, who produced the illustrations for this edition; and to the Director, Princes Risborough Laboratory of the Building Research Establishment for the loan of much of the material from which the drawings were made.

DGP

Part One

INTRODUCTORY

1 Structure and properties of wood

Wood has now been used by man, even if only as a fuel, ever since he has needed warmth and shelter. In relatively recent years the range of uses for wood and wood products has extended enormously – we live our daily lives surrounded by it in one form or another. This popularity is partly due to its ready availability in virtually all countries; it is also due to the fact that trees replace themselves – timber is a renewable resource, especially so when man takes care to replant trees to keep pace with those he has cut down, or at least allows time for new ones to regenerate naturally. There is no reason for a shortage of timber to develop in the world. However, man's insistence on using certain 'fashionable' species in greater quantity than can be replaced naturally has led to the disappearance or at least rarity of a number of once well-known species. Scarcity also leads to increased prices.

Several once little-known timbers, such as Virola, are now used on a larger scale and undoubtedly in the future we are going to be finding ourselves using species that we scarcely know now. The properties of timbers vary but many species can be interchanged with one another for a particular use and the performance of the finished article will not suffer. Appearance can be altered with the use of stains and finishes so this is no difficulty either. In recent years there has been a marked shift away from solid wood (especially in the case of hardwoods) and the use of wood veneers or plastic surfaces on particle boards has become common. Often the artificial surfaces are patterned to resemble real wood; if they are textured as well it can be quite difficult to distinguish them from the real thing.

3

One of the most recent changes has been the development of a fibreboard made with adhesive; this is known as MDF (medium density fibreboard) and it can be machined and used almost as though it were solid wood. It can be stained, painted or veneered and is widely used in furniture. Before long it may also be used in joinery on a large scale.

It is sometimes said that these other materials are inferior and cheap but this is unfair; no one would suggest that a cake is inferior to the butter and sugar that went into it – it is simply different. It may be that the new materials are cheaper but this is not always so. It certainly isn't true to say that they are inferior either. The various sheet materials like plywoods and fibreboards have been developed because the particular properties that they offer are valuable, desirable and cannot be obtained from solid wood – in particular this is true of size and stability. A material is only inferior when man mistakenly uses a grade or type that is not really suitable for his particular purpose. This may happen through ignorance or meanness but is not the fault of the material. Many sheet materials are very strong and all are available in much larger pieces than natural solid converted timber – this makes their use in manufactured items highly desirable for many designs.

Trees grow in all countries and provided there are enough of them they can be exported to other countries. Sheet materials have added to this trade from the 1930s onwards and are now very prominent. In more recent years the movement of whole logs has all but ceased, because most exporting countries prefer to do the first stages of conversion themselves and so extend their own industry and employment opportunities.

The shipment of blanks, semi- and fully-machined parts is now gathering momentum and trade in fully-finished items, such as hardwood doors from the Far East, is steadily increasing.

Timber has become ever more widely and easily used in the 'Do-it-Yourself' market and a vast range of developed products is available for the use of householders, architects, builders and other manufacturers. The ready availability of simple wood-working machinery, treatments and finishes continues to add to the appeal of wood.

Many users think of wood as a single product, but there are hundreds of different types and they exhibit a wide range of properties. While many are quite safely interchangeable this is far from universal – a timber for a particular purpose needs to be carefully selected to get the best combination of properties available in a given price range.

For identification purposes, reliance is placed on the most obvious factors such as the weight, colour, smell and figure (which is usually incorrectly called the 'grain'). These factors are useful ones, but are not necessarily reliable or constant, and in some cases (as, for example, when the timber has already been incorporated into a building, or has been treated with a preservative or staining agent that hides its structure) it may not be possible to identify the wood with any certainty at all.

Under conditions such as those just described it is an easy matter for an unscrupulous dealer to pass off a cheap, inferior wood as a greatly superior variety, and in constructional work a considerable element of risk may be introduced by the use of an inappropriate timber. Even with no deliberate intention to defraud, mistakes are possible that may lead to mechanical failure of the timber whilst in use. In his own and in his clients' interests every dealer, builder and architect should be able to check the species of timber supplied to an order. Unfortunately this can often be done only by examination of the wood with a microscope or hand lens, but such a test is possible even when only a small sliver can be taken for study. Those features of the wood that are visible to the naked eye or under a hand lens giving low magnification are described as the 'macroscopic' features, whilst those for which a microscope is needed are termed the 'microscopic' features. In this book emphasis has been laid on identification by macroscopic characteristics; checking a timber's identity by this means is possible without any extensive training.

For many of the common or well-known species included in this book small drawings of end-grain structure are provided; these will be explained in a later section.

Factors that affect the correct utilization of wood, such as preservative treatments and drying are in general well understood by timber users, but are included in later sections to provide a more complete picture.

GROWTH AND STRUCTURE

When a seed of any tree species germinates, it sends up a shoot which may attain only a few millimetres in height or may reach a metre, depending on the species, soil and climate. This shoot consists of a central pith of very thin-walled cells (cells are the building units that all living things are made of). Around the pith is a layer of tissue composed of cells which perform various functions: principally strength, storage and transport of foodstuffs. The layer is referred to as the first growth ring and can be seen if the shoot is cut to show its cross-section.

On the outside of the growth ring is a very thin sheet of specialized cells called the cambium, and outside this is the inner bark layer. The cambium is almost too thin to see but it is very important to the plant because it is where growth takes place every season to form a new growth ring. Each cambium cell divides when it has grown to its full size; one new cell regrows to a mature cambium cell ready to divide again and the other new cell either forms part of the new growth ring or new inner bark layer – both new layers are formed each season. The gradual accumulation of many growth rings over the years leads to the formation of a thick woody stem which we call a log when it has been cut down. The most recently formed growth rings conduct water and mineral salts from the soil up the tree to the crown of branches; in the leaves these materials are con-verted to foodstuffs which pass down the tree in the sap of the inner bark. The cambium is well supplied with food so that growth is pos-sible. As the trunk grows thicker inside it, the inner bark becomes stretched and so is constantly replaced with a new layer each season; the previous season's layer of inner bark simply dies and becomes the protective outer bark. The outside surface of the bark slowly flakes off so that the bark does not build up to more than a certain thickness.

The shoot grows upwards each season as well as growing thicker, so that a season's growth is really a cone. This means that a tree is slightly tapered – if the taper is severe it can give problems to a sawmiller.

The cells in a growth layer or ring have tiny holes called pits in their walls so that foodstuffs can move freely. Only a few of the outer growth rings take an active part in the conducting of food; this zone is known as the sapwood and is always pale in colour. The cells in the sapwood are fully grown and have normal strength properties. As new growth rings form, the cells of the innermost growth rings are eventually used for the permanent storage of waste products produced during utilization of food materials for growth. These waste products may be colourless or may be some shade of yellow, pink, brown or black and the zone containing them is known as the heart-wood. If the colour is an obvious one then the central zone of a trunk or log will be said to have a visible heartwood; examples would be Afara, Light Red Meranti, Yew or Ebony. If no colour is obvious then the heartwood is indistinguishable from the sapwood as in Birch, Beech or Spruce.

Very often the waste products give the tree protection against fungal decay – the durability of the cut timber depends on how toxic to fungi the waste is. In general it is true to say that the darker the heartwood is the better the durability will be; sapwood never has any natural durability so will always need preservative if used in an exposed situation.

In the first few years of growth the tree is all sapwood; once the heartwood starts to form it will keep pace with the production of new layers of sapwood – most trees have about 50 mm of sapwood and this is perfectly safe to use unless one is concerned with colour or durability. A large log obviously will have a much greater volume of heartwood.

In some species deposits of a variety of chemicals may be found. For instance tiny invisible crystals of silica occur in a few timbers and these blunt cutting tools quite rapidly. Other timbers can have gums or resins – these also affect machining because of their stickiness. Some substances may produce a characteristic odour or colour.

There may be a marked difference in the type of growth shown by a growth ring during different periods of any one growing season. This has given rise to the expressions 'earlywood' and 'latewood' to denote wood that has grown in the first part or later on in a growing season. The rate of growth (the width of growth rings as seen on the end of a log) can vary quite a lot with soil and climatic conditions; this can make quite a difference to the weight, texture and other properties.

Softwoods and hardwoods

All commercial timbers are divided into two classes: the softwoods and the hardwoods. This classification has nothing whatever to do with the weight or hardness of the wood concerned, but is based solely on its structure – Balsa, weighing on average only 160 kg/cubic metre, is easier to work by hand tools and has less surface hardness than Pitch Pine, which may weigh about 720 kg/cubic metre, but Balsa is classified as a hardwood and Pitch Pine as a softwood.

Botanically, trees are divided into two groups: the Angiosperms and the Gymnosperms.

The Angiosperm group contains all the trees that provide the hard-woods, and are those trees whose seeds are carried in a seed case or in fruit; such trees bear true flowers. The group is sub-divided into (a) the Monocotyledons, the plants having a single seed leaf and (b) the Dicotyledons, those plants with two rudimentary leaves in the seed embryo. In general, commercial hardwood timbers are provided by the Dicotyledons. Those growing in temperate and cold regions are mostly deciduous. In warmer and tropical regions most hardwood species are evergreen.

The Gymnosperms are the conifer (cone-bearing) trees and are usually evergreen; this group produces all the softwoods.

The two types have rather different general structures and uses. A brief description of each type follows to help in their identification and use.

Softwoods

A tree is composed of billions of cells; they are so small that they cannot be seen with the naked eye, though it is possible to study individual cells under a microscope if they are separated out by a pulping process. In the softwoods these tiny cells are shaped like tubes that have closed ends which are somewhat pointed. Cell walls are always made from a mixture of stretchy cellulose and rigid lignin. The proportion of these two substances, and therefore the strength of the cell varies in different cells and species of wood. Softwood cells have pits for food circulation, and the function of the cell is to conduct food and provide the mechanical rigidity of the tree at one and the same time. In such timbers these cells are called tracheids, and they are packed closely together parallel to the length of the tree or log. In the spring, cells with thin walls but relatively large cavities are formed (the earlywood, used for conduction), whilst in the summer the cells have thicker walls with correspondingly smaller cavities (the latewood, used for giving strength to the tree). It is the noticeable difference between the earlywood and the latewood in many softwood timbers that defines the growth rings so well, and gives the timber the appearance of being built up of layers of wood when a log is examined in end-section.

The soft tissue responsible for the storage of food in the growing tree (as opposed to its conduction) is known as the parenchyma, and may be of two kinds, either wood parenchyma or ray parenchyma.

Wood parenchyma cells are more or less brick-shaped and form strands running along the grain. These strands are rare and hard to see in softwoods.

The ray parenchyma again consists of oblong-shaped cells that have thin walls with pits. This tissue forms very narrow, light lines that may be visible on the end-section of the log, running from the centre of the tree to the bark. These bands of ray cells (called simply rays) do not necessarily start at the middle of the log, but each one continues outwards until it reaches the inner bark. Rays are responsible for sideways movement of food materials and waste products. Although present in softwoods, the rays and parenchyma are not of such importance in identification as they are in the hardwoods.

Many softwoods contain resin canals that are visible under a hand lens either as small brown-coloured dots or tiny openings on the end-section of the wood. The resin canal consists of an actual hole that is surrounded by special cells called epithelial cells that secrete resin into the hole. The canal is lined by a complete sheath of epithelial cells. These resin canals may be empty in old dry wood.

Pitch pockets, which are longitudinal pockets of resin caused by some injury to the cambial tissue during the growth of the tree, are sometimes to be found in the softwood species.

Hardwoods

In the hardwoods different types of cell do the separate work of conducting food, giving structural rigidity, and storing food. Those that conduct food are called vessels (or pores when looked at on an end-section under a hand lens); they resemble small round or oval holes in the wood. Each individual vessel is a tube that extends for many metres along the trunk or branch and is made from the joining together of many cells in a vertical line. In the early stages each cell is completely closed, but later it swells out into a cylindrical shape and opens at both top and bottom to join its neighbours above and below. Each vessel is, in effect, a hollow tube, and a series of vessels form a circulatory system from roots to leaves of the tree, the conduction of the food being helped by pits in the vessel similar to those in the tracheids of the softwoods. The end walls of the vessel members may be either completely absorbed at the time the vessel is ready to start functioning (in which case the vessels are said to have simple perforations) or else be incompletely absorbed so that thin bars of tissue connect the side walls of the vessel member; this last type of vessel perforation is described as scalariform. It can be seen as a tiny ladder-like structure in a vessel line on a quartered surface.

Apart from the perforations, the vessels are a good diagnostic feature according to their manner of distribution when the wood is viewed in end-section. They may be scattered about the wood as single vessels (as in Horse Chestnut) when they are called solitary vessels; in radial groups or lines, either long or short, (the term radial meaning to cross the growth rings in the same way as do the rays); or in clusters, as opposed to the radial groups. Radial lines or groups of vessels are normally made up of a certain specified number of vessels, the number varying according to the species of wood. In some timbers, a combination of radial lines or groups and solitary vessels is to be seen, as for example in Lime, where the solitary vessels and the short radial groups can be clearly distinguished. Chains and groups of pores may also run parallel to the growth rings (tangentially) or may tend to run diagonally to the general direction of the rays, an arrangement that is described as 'oblique'. Usually it is only possible to see such detail of arrangement clearly by using a hand lens and, obviously, learning to recognize the many variations in structure is a matter of study and experience. The drawings given for many species in this edition are intended to help the reader to recognize the more common timbers and check whether a piece is what they think it is.

The size of the pores in the early- and latewood of certain species also serves as a useful feature for identification purposes. In some timbers (for example Ash, Alder, Oak and Mulberry) the vessels of the earlywood are much larger than those of the latewood, with an abrupt change in size between the two classes, the larger vessels form-ing a more or less continuous ring to mark the boundary of the growth ring; such timbers are called ring-porous woods. These timbers are invariably resilient and well adapted to resist suddenly applied loads, and are therefore well suited for use as tool handles and similar items. Very few tropical hardwood timbers show a ring-porous structure. Most hardwood timbers (e.g. Maple, Sycamore, Pearwood and Birch) do not show this abrupt change in vessel sizes; woods that show the end-sections of the vessels in a growth ring to be all much the same size or only gradually decreasing in size are said to show diffuse-porous structure.

With most dried timbers the vessels in the sapwood are empty, having given up their food content during the process of drying, but in the heartwood the vessels may have become filled with solid or semi-solid deposits of waste. It is sometimes possible for these deposits to cause staining of any material in contact with the wood under damp or wet conditions. An example of this is Afzelia – a durable timber suitable for window frames and other external joinery. Rain will leach out yellow deposits which then stain brick or rendering nearby.

In addition, tyloses (or, in the singular, a tylosis) may be formed in these vessels. These are minute bubble-like structures formed from the pits when pressure changes happen as sapwood is converted to heartwood. Tyloses may form to such an extent that they completely block the inside of the cell. Under examination with a hand lens the tyloses show up as small glistening spheres, giving the general impression of minute bubbles. Timbers that have such tyloses are difficult to treat with wood preservatives, as the tyloses form barriers that prevent easy impregnation by preservative fluids. Mulberry, Oak, Ash and Teak are four quite common timbers in which tyloses are always to be seen in the heartwood.

On the two longitudinal surfaces of a piece of wood, known as the radial surface when cut parallel to the rays, and as the tangential surface when cut right across the log (or, in other words, the quarter-sawn and flat- or plain-sawn surfaces respectively in trade nomenclature), the longitudinal lines made by the columns of vessels may sometimes be seen either with the naked eye or with the help of a hand lens. These lines are called vessel lines, and they may also serve as an identification feature. For instance, vessel lines are visible only with a lens in Pearwood, they are indistinct in West Indian Satinwood, quite distinct in Mulberry, and in the European Oak are so distinct that they appear to the naked eye almost like coarse scratches. The vessel lines are sometimes thrown into greater prominence by the deposits in the vessels, as in the case of Cherry.

The cells that form the strengthening tissue of a hardwood are shorter than the tracheids to be found in the softwoods, and their ends are rather more pointed than those of the tracheids. They are called fibres in the hardwoods and the strength properties of a timber depend largely on the percentage of such fibres to be found in it.

As in the softwoods, the storage tissue of hardwoods consists of parenchyma of two kinds, wood and ray, and either or both may provide good features to aid identification. Wood parenchyma occurs to some extent in most hardwoods and is often visible as some sort of pattern on the smoothly cut end-grain. It may be just single cells (diffuse parenchyma as in Mersawa) that are present and then it is almost impossible to see without a microscope. In many timbers, fortunately, the storage parenchyma is easily visible as layers or lines marking the growth ring boundaries (terminal parenchyma, as in American Mahogany); or as more frequent layers or bands running round the growth rings (as in the fine lines of Walnut or the thicker bands of Guarea). Parenchyma can often be found associated with the vessels; vasicentric parenchyma is a thin layer around the vessel forming a sheath, as in Greenheart; if parenchyma around vessels is drawn out into points sideways like wings it is called aliform – this form is present in Afara and Ramin. Sometimes the points join up in an irregular way – this type is referred to as confluent and a good example is Iroko.

The second type of storage tissue is the ray parenchyma; the rays of hardwoods are more in evidence than they are in the softwoods and are also far more important as an identification feature.

When ray tissue is laid on with a width of one cell only, the rays are described as uniseriate (as in Poplar) whilst rays two or more cells broad are described as multiseriate (as in Birch). Occasionally ray tissue is to be found consisting of a large collection of rays with a few fibres scattered through the ray tissue. Such an arrangement is described as an aggregate ray, and rays of this nature are found in Hornbeam. This last type is not common.

In some woods the rays are to be seen longitudinally (on the tangential face) in more or less uniform rows of the same height, the rays being spaced very finely and usually indistinctly except with the help of a lens. Such an arrangement is very noticeable in American Mahogany. The name of ripple marks has been given to this storeyed structure of the rays in hardwoods, and the presence or absence of them may be an important factor when attempting to identify a timber. Commercially, the rays are important in that timber sawn radially (quarter-sawn) shows the rays making an attractive figure across the wood, as with the so-called 'silver grain' of Oak. Only quite large rays are obvious in this way.

Other features of importance in the hardwoods are the resin canals and pitch pockets briefly considered in the remarks on the softwoods above, but in the hardwoods these are only occasionally to be found. In a few timbers they may be found horizontally in the ray tissue or vertically on the end-grain.

Pith flecks, which appear as small brown lines running in the direction of the grain, are caused by damage to the cambial tissue by the larvae of a species of small beetle. Such pith flecks are very strongly marked in Birch and are also to be seen in Poplar. They normally occur only in certain species.

A very few tropical hardwoods also show latex (liquid rubber) canals and in some woods they may be very long in radial section and be of such frequency as to render the wood unsuited for structural use except in very small lengths. Economically, the importance of such timbers rests on the products that can be obtained from their latex as, for example, guttapercha.

NAMING AND IDENTIFICATION OF TIMBER

The commercial naming of timbers presents many problems. For instance, what differences, if any, are there between French, Spanish, English, Persian and Circassian Walnut? Do an Englishman and an American mean precisely the same timber when they speak of a piece of Ash? Is Red Fir the same timber to all intents and purposes as Yellow Fir, and if so, why is it that both may be referred to as Northern Pine or Baltic Redwood?

The commercial nomenclature of timbers is both bewildering and unreliable, more especially in the naming of tropical woods, for in the past the custom has been to describe a little-known timber by reference to some better-known wood, plus the addition of a geographic name, as for example Rhodesian Teak, which is not a true Teak at all. The names Mahogany, Oak, Walnut and Maple, in particular, prefixed by the name of a country, district or colour, have been favourite ones for popularizing local timbers to build up an export market, especially where the local timber bears some superficial resemblance to the true wood. In some instances no great harm is done but this practice can easily lend itself to the passing off of inferior timbers as much superior varieties, when the inferior wood may have none of the characteristics of the timber after which it has been named and may perhaps be entirely unsuitable for use under conditions in which the true timber would be invaluable. Certain names of this type are well established but the Trades Description Act will make sure that no more happen. A wood can only be described as Teak or Oak if it genuinely is – otherwise it would be referred to as teak or oak type, or finish. In more recent years the trade has tended to make use of local trade names rather than borrowed titles, which is a step in the right direction.

Unfortunately this trend has led to an additional difficulty being created by having a common local name used to describe entirely different timbers in other parts of the world. In Africa a certain timber is named Yellowwood, whilst in Australia the same name is given to a totally different wood, a confusion that is made worse by the fact that one timber is a botanical softwood and the other a hardwood. Additionally, in any one country, especially large countries, there may well be more than one common name available because there

13

is more than one language in use in different areas. This is especially true for many African and Asian countries.

We therefore have a situation where one timber may have more than one name in a given country and it also grows in several countries; an example of this is Iroko, which has been known in this country by at least seventeen different names.

In order to try to rationalize the position the British Standards Institution has produced Standard 881 & 589 which is *Nomenclature of commercial timbers, including sources of supply*. This publication lists all likely names for a species but proposes a standard name by which it should be known in the trade regardless of the country it was imported from. In general this system is followed fairly well and there is much less confusion than there used to be.

Scientists do not have this difficulty of nomenclature because they use botanical names. Plants have been carefully divided into families according to their characteristics, and sub-divided into genera; each plant is described by a two-part name that gives the genus followed by some specific description, the whole being in Latin. The specific name may be a proper name, commemorating a country or person who first found it, or it may be descriptive of the colour, type of leaves or some unusual feature such as having spines on its surface. As an example, take four timbers known commercially as American White Oak, American Red Oak, Turkey Oak and Evergreen Oak. These would be described by a botanist as *Quercus alba*, *Quercus rubra*, *Quercus cerris* and *Quercus ilex* respectively, the names indicating that they belong to a common genus and share certain characteristics. Reverting to an example previously quoted, the botanical name of Rhodesian Teak is *Baikiaea plurijuga*, which shows that the timber is not of the same genus as that of the true Teak (*Tectona grandis*) and that consequently the two timbers may have quite dissimilar properties.

It is not possible (or perhaps even desirable) to try to substitute a botanical system for the commercial nomenclature of timbers, but with the aid of the former it is possible to verify and compare laboratory tests on timber that may have been carried out in widely scattered parts of the world without the confusion that is very likely to arise through the use of local or trade names. The practice of including the botanical name in brackets after the trade name, when drawing up a specification, is gradually increasing and will do much, if generally adopted, not only to check possible attempts at fraud or negligence, but also to ensure that timbers are more correctly utilized. In Part

Two of this book botanical names have been given, as well as the better-known alternative commercial names. With the help of these it is possible to select alternative timbers for specific purposes. Two timbers with the same generic name are likely to have similar properties and this makes a good starting point.

The identification of timber

We may now consider the study of a piece of timber with a view to its identification.

The timber to be examined is cut across the transverse section (end-grain), using a very sharp knife and making a single clean cut along the rays from the direction of the outside of the wood towards the pith or centre. The structure can then be examined closely with the aid of a hand lens that gives a magnification of about ten times. Such important items as the relative size of the vessels, arrangement of the vessels either in ring-porous or diffuse-porous formation, groups, chains or solitary vessels, presence or absence of gum ducts, vessel deposits, tyloses, and the type or types of wood parenchyma present are all important points and should be carefully noted. References to the appearance of vessel lines and the presence of absence of ripple marks on longitudinal surfaces are also useful.

After examining a specimen, a description of the timber along the following lines might be available.

Ring-porous structure. Very distinct growth rings, the boundaries being marked by large vessels and layers of terminal parenchyma. Vessels oval in end-section and variable in size; some solitary and some in short radial groups. Vessel lines coarse and quite distinct to the naked eye. A certain amount of diffuse wood parenchyma present, but considerably more vasicentric parenchyma. No ripple marks to be seen, nor any pith flecks.

The above is not a complete description of the structure as it could be revealed, but is sufficient to show the points that should be fairly readily noted. The total pattern of structure as seen on the end-grain is frequently particular to a given genus or species and acts as a means of identification, much as the fingerprint of a human being. In a few cases very closely-related species cannot be distinguished from one another even when a microscope is available to see more detail. In most cases, however, the structure is fairly distinctive – the only trouble then is to recognize a particular overall pattern and to give a name to it. We are quite used to doing this with people's facial characteristics and do not often fail to recognize people we know; police have more trouble with fingerprints and use specialists to sort them out. In the case of timber it is difficult to become familiar enough with a large range of structures to be able to recognize them with certainty. The more expert you become the more you are aware of the possibility of mistakes.

In this edition we have included small end-grain sketches of the structure of about a hundred of the more common hardwoods. Unfortunately, softwoods all look rather similar (earlywood and latewood, narrow rays, no parenchyma, resin ducts in just a few species) and drawings are not really helpful. The intention of the drawings is to enable the reader to make a comparison of the structure they find in a sample with that of the standard drawing and in this way they can often confirm an identification. Remember that some slight variation is always possible with biological material so you may not always get an exact match, just as people are not all identical although they are the same species.

Where the identity of a timber is suspected prior to its being examined, it is usually a simple enough matter to check the structure with the data published for that timber and so prove or disprove its identity; where the identity is in doubt from the outset, reference to a key is necessary unless you go all through the pictures hoping for a match.

The principle underlying what is called a dichotomous key can be understood quite easily, for identification is arrived at by a process of elimination by working through a series of questions, the questions being arranged in pairs.

As a simple example let us suppose that the first group of two questions calls for an answer as to whether vessels are present or absent in the wood under examination. There can be only a straight-forward 'yes' or 'no' and the wood is identified as a hardwood if the vessels are present, or as a softwood if they are lacking. Each section of the key carries a reference to a later section. Suppose that it has been decided in our imaginary examination that vessels were present in the specimen. After the words 'vessels present' in section 1 of the key, '(see section 7)' might appear. On reference to that section we are faced with the alternatives '(a) Ripple marks present (see section 29)', and '(b) Ripple marks not present (see section 32)'. Assuming that ripple marks were visible, we would pass on to section

29, but as can be seen, by that time we would already have eliminated from consideration all the softwoods and all the hardwoods that do not show ripple marks. The process of elimination would then be continued until the final question indicated the identity of the timber being examined.

We can take a very simple example of how this type of key works, assuming that we have a piece of Sandalwood (*Santalum album*) that we wish to identify. We must assume that the identity of the timber is unsuspected, but that it is known to be one of a group of ten timbers that includes Sandalwood. The other nine timbers are common Ash (*Fraxinus excelsior*); Boxwood (*Buxus sempervirens*); Deodar (*Cedrus deodara*); Yew (*Taxus* species); Honduras Mahogany (*Swietenia macrophylla*); Teak (*Tectona grandis*); Honduras Rosewood (*Dalbergia stevensonii*); English Oak (*Quercus robur*); and Douglas Fir (*Pseudotsuga menziesii*). The number of timbers has been limited to keep the key easy to explain.

Turning to section 1 of the key covering these ten timbers we find:

1 ... (a) Vessels present? (see section 2)
   (b) Vessels not present? (see section 3)

Examination of the structure shows that vessels are present in the specimen, so the softwoods Deodar, Yew and Douglas Fir are automatically eliminated from consideration. We pass on to section 2, which may read:

2 ... (a) Ring-porous or tendency to ring-porous structure present? (see section 5)
   (b) No tendency towards ring-porous structure? (see section 7)

Our sample shows a positive diffuse-porous structure, thus eliminating Ash, Teak, Oak and Honduras Rosewood, the three former timbers all being most definitely ring-porous in structure, whilst Honduras Rosewood shows a tendency towards being ring-porous.

Section 7 offers us:

7 ... (a) Ripple marks present? (see section 11)
   (b) Ripple marks not present? (see section 15)

Ripple marks are not present in the specimen we are examining,

17

and Honduras Mahogany is thereby eliminated. Passing on to section 15 we would find (for this particular key):

15 ... (a) Occasional tyloses present? ... Sandalwood
    (b) Tyloses not present? ... Boxwood.

As our sample has a few tyloses, it is identified as being Sandalwood.

It is not pretended that the above is a full key, or even the best for the ten timbers concerned. For instance, no account has been taken of the arrangement of vessels, vessel lines, gum deposits, rays, etc., but it has served as an example of a dichotomous key and it will be noted that such items as weight, colour, etc., that might be very variable, have had nothing to do with the identification which has been based on structure alone.

With the hundreds of commercial timbers available, no completely satisfactory key of this nature can be published that will cover them all, and each worker needs to prepare his own key, incorporating in it only those woods that he is likely to encounter in his professional capacity. Dichotomous keys for various groups of timbers are however included in various publications and do not require too much scientific knowledge to be easily used.

For the person who may need to identify some hundreds of different timbers the dichotomous key is not as valuable as the Punched Card Key. This consists of a series of cards of standard size, one card being prepared for each timber. Round the edges of the cards various feature headings are printed in an unvarying order and each heading has a hole punched near to the edge of the card. Each timber to be included has a named card made up with a series of punch holes cut open to the edge of the card according to the features present in that timber. Each card is therefore a concise description of that named timber. Identification is made by passing a long needle through the total set of cards using a hole which corresponds to a feature observed in the unknown specimen and lifting the needle a little. All the cards notched open for that particular feature will fall from the needle and those remaining on it can be discarded. The pile of cards diminishes as each feature observed in the specimen is checked through, until finally only one card is left which includes all the features observed; this card bears the name of the timber being identified. Suitably printed cards for this type of key are readily available. For those wishing to prepare such a key, reference should be made to *Identification of Hardwoods: A Lens Key* (Forest Products Research

Bulletin No. 25, second edition, 1960, HMSO), which gives identification features for a great number of timbers, together with full details of how such a key can be prepared and handled. It is available through libraries only since it is now out of print. Cards are available through Her Majesty's Stationery Office.

Before any attempt is made to classify timbers of which the identity is doubtful, some practice should be obtained with known timbers, the structure seen under the lens being compared with the printed data so that the various types of parenchyma, vessel arrangement and so on can be recognized on future occasions. We have all spoken to someone who we thought we had recognized and felt embarrassed when we found we were wrong; there are hundreds of commercial timbers and we cannot know them all, so be cautious with recognition. Large trade suppliers are very careful and not too often wrong.

Selection of the right timber for a particular job relies upon a selection of appropriate properties. Timber of a different species may well have different properties so identification checks are obviously very important.

GRAIN AND FIGURE

This section is to do with the appearance, finish and texture of a wood surface and it is best to give precise definitions of terms in common use. Many of these terms are often misused.

Grain – The word grain, more than any other, produces continual misunderstanding owing to the looseness with which it is used. For the purpose of the study of timbers grain has nothing whatsoever to do with the appearance of the wood and means merely the direction of the fibres relative to the axis of the longitudinal surface of the wood under examination. Wood splits along its grain quite easily.

Straight grain – indicates that the fibres run parallel to the vertical edges of the tree. This gives maximum strength.

Diagonal grain – has fibres inclined to the edge of the wood. This may be a serious defect (as it has a considerable weakening effect) and is brought about by sawing the wood diagonally in the timber yard, or because the grain spirals in the original log.

Spiral grain – is where the fibres occur either in clockwise or anti-clockwise spirals (another source of structural weakness).

19

Irregular or cross-grain – refers to the deflection of grain round knots, or the way grain slopes through a piece.

Interlocked grain – where successive layers of fibres are inclined in different directions the grain is said to be interlocked; timbers showing this feature are difficult to work with hand tools and exceedingly difficult to split radially. It is sometimes called alternating spiral grain. Interlocked irregular grain produces what is known as roe figure. Timbers having an interlocked grain show a banded or striped figure on quarter-sawn stock, the striping varying according to the degree of interlock present.

Wavy grain – causes the popular 'fiddle-back' figure.

Figure – may be defined as the pattern on the wood; it is produced by such things as the colour difference between the early- and the latewood, the pattern made by the growth rings as seen on the longitudinal surfaces and the actual type of grain present. The amount of ray material present will also affect the figure.

Texture – is another word that is frequently misused but when used correctly it refers to the prevailing size of the cell cavities. Greenheart is said to have a fine texture because the vessels are small on end-section, whilst Oak is said to be of coarse texture because its cell cavities are relatively large. Texture may range from very fine to very coarse according to the species of wood. It may also be described as 'even' (or 'uniform') or 'uneven', the difference being that uneven texture is to be seen in woods in which there is a marked contrast between the early- and latewood (as in the ring-porous woods), while where no such difference exists the texture is said to be even, or uniform. Texture can be chosen to a certain extent, but sometimes it is necessary to use a coarse textured wood and also to polish or paint it. In such a case it is necessary to use a filler – a material which fills the vessel lines and so gives a smooth surface for the application of finishes.

COLOUR AND WEIGHT

Very few timbers have a really distinctive colour. The black of Ebony, stripes of Zebrawood and bright irregular colours of Rio Rosewood are not too common. Most species can be described as creamish or brownish and brown might be a red-brown or a gold-brown.

Sapwood is always very pale; heartwood in some cases is the same colour as the sapwood (Beech, Birch, Sycamore, Obeche, etc.) and the whole log is used for timber. Many species have a darker heartwood and this is what we usually mean by colour. Some figures show up not because of actual colour variations but because leaning fibres (interlocked or wavy grain) cause a variation in reflectivity.

Some species are a bit variable in colour and care has to be taken that successive consignments will match each other. Colours may also fade or darken on exposure to bright light and especially direct sunshine, so a new piece will not match an old piece until it has changed with the passage of time. The use of stains or pigmented finishes on wood goes a long way to get over this problem and probably arose because of it. Nothing looks as good as a natural colour and finish and a little variation is usually acceptable.

The weight of any given species of timber will vary according to the conditions under which it was grown, the soil, its position in the tree, age and many other factors. There may be a fourfold variation in weight for different samples of the same species although variations this large are not common. The weight figure quoted in the entries on the different timbers may be taken as a fair average weight for that timber. All weights are given for timber in the dry state, that is to say when dried to the range 12–15 per cent as would be required for indoor use. Softwoods vary from about 350–700 kg/m³ but hardwoods are much more variable, 200–1200 kg/m³ being the general range. When selecting a species for a particular mechanical use, weight is obviously more important than appearance because strength is connected to it. Appearance can always be altered; a need for a red or blue timber can be satisfied with an application of stain. Strength is built-in very largely as a result of weight; it can only be increased by increasing size of section but this may make the weight unacceptable, especially in the case of furniture.

Weight is the best single indication of likely strength; when it is necessary to use an alternative timber for a job then species of similar weight should be looked at first.

When durability is an important requirement for a given purpose then colour is a useful indicator; in general pale timbers are perishable and darker timbers are durable. There are of course exceptions to this – it is not an invariable rule.

DEFECTS

Trees can suffer from certain difficulties during their lives which can give rise to abnormalities in the timber. Some of these are so noticeable that they are removed during sawmilling and conversion – others can be difficult to see until failure occurs while the wood is in use. Three fairly common defects are listed here.

Natural compression failures

This defect is more widely, if incorrectly, known as 'thunder shake'. It can occur in any timber throughout the world and consists of a fine hair-line crack across the grain which fails under load. It can develop in the standing tree through storm damage or on felling as the trunk hits the ground. It is caused by localized over-compression. Apart from the crack (which is only visible after planing) the rest of the timber is unaffected and is safe to use.

Brittleheart

This defect occurs in light to medium weight tropical and sub-tropical hardwoods which have passed their prime and become overmature. Their great weight causes fibres in the central zone of the standing tree to crack across – this throws extra weight onto the neighbouring fibres and so the crack gradually spreads. A large number of such cracks will develop and gradually the zone containing them will spread up the tree and outwards, forming an ever increasing cone shape of brittle timber in the heart. In severe cases over half the volume of the log may be affected. When converted such timber may break during sawing – if not it may break later when the wood is in use.

The difference between this and thundershake is that the length of wood is all affected, even where no crack is actually visible; cracks will develop later under load. In the early days of timber trading such affected material was called a '3-man board' because unless a third man carried the middle of a board it might break under its own weight. Smaller logs are much less likely to show this defect; if it is suspected then a knife test may help – stick the point of a penknife a little way into the side grain at an angle and gently try to lever up a small splinter. If brittleheart is present the splinter will break over the knife point; if the wood is sound it will almost always provide a longer splinter that lifts up at one end.

Reaction wood

If a tree is growing on a slope, subject to a prevailing wind or leaning over for any reason it will develop an asymmetric trunk. The cross-section will tend to be oval rather than round and the pith will be off centre, perhaps by a long way.

If the tree is a softwood the larger radius will be on the underside of the lean and is therefore called compression wood. This radius will contain more lignin than usual and be very hard; during drying much of the wood in this radius will shrink badly along the grain and so cause much warping.

If the tree is a hardwood the longer radius will be on the upper side of the lean and is called tension wood. This longer radius contains more cellulose than is normal and so the wood is softer – sometimes almost rubbery. It will be difficult to machine smoothly and sawing it green will produce very fluffy surfaces. Again, shrinkage along the grain is severe and so is the resulting distortion and possibility of movement in use.

Both of these defects are hard to see once the oval log has been cut; both are responsible for quite a lot of the bad behaviour found in some pieces of wood.

2 Wood: its preparation for use

CONVERSION AND MACHINING

Before it can be used for anything, except perhaps as a fuel, wood has to be cut into suitable sizes. Primary conversion is the cutting of a log into large planks or sections ready for the drying process. Cutting dried wood into specific smaller sizes or pieces is known as secondary conversion.

The simplest and cheapest system of primary conversion is to cut a log with a number of parallel cuts along the grain (see below). This is known as cutting 'through and through'. Most of the boards have a tangential face – that is the growth rings are going more or less across the board. Such boards are called flat- or plain-sawn. The central boards are called quarter-sawn and have radial faces – that is the rays are going more or less across each board.

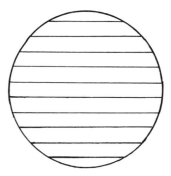

Another method is to cut the log into quarters first and then cut each quarter with cuts which are more or less all parallel to the rays. This provides a total supply of quarter-sawn boards but requires more handling and therefore time and money.

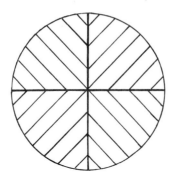

A third method is called plain sawing and this gives a total supply of tangential boards. The central piece is used for unimportant purposes because it is where defects are most likely to be. Again a fair amount of handling is required so it is more expensive than sawing through and through.

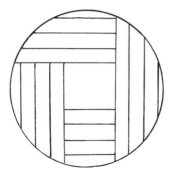

At this point it is necessary to explain that wood is hygroscopic – that is, it is able to absorb water from damp air or give up moisture to dry air. As the moisture content of the wood changes it is liable to change in shape as well – the extent to which a wood is prone to this is known as its stability. Quarter-sawn material is much more stable than flat-sawn and is therefore preferred for certain purposes, particularly furniture. This is the reason for different methods of primary conversion but the customer must be prepared to pay more for the method which involves the most man-hours: the quarter-sawing.

For general purposes a great deal of wood is simply sawn through and through – the most economical and least wasteful method.

After drying, wood is further converted into a large range of sizes for construction or any other purpose. Many special profiles are standard production, whether you require larger sizes for window sills, or smaller ones for applied beading or strips.

Some parts of the world use imperial measurements and others use the metric system. Since the timber trade is international it uses both. The result is a series of awkward metric sizes that are simply a conversion of an imperial size and not something simple like 200 mm × 50 mm. It is not unusual to find sizes whose sections are measured in millimetres being sold by the foot of length so care needs to be taken with orders and specifications to avoid misunderstandings.

Machining is the term generally applied to processes that cut or smooth to achieve a precise size and finish, cut special shapes and profiles, make joints, holes or whatever other detail is needed to manufacture something. Wood is an easy material to work with and this accounts for its popularity in the Do-It-Yourself field as well as in professional craft and construction.

A few timber species contain mineral crystals that can be very blunting to cutting edges, or gums that produce an annoying amount of stickiness, or have a grain direction that causes 'picking-up' or roughness of a finished surface. A particular problem arises with machining knotty timber because of the hardness of the knot and the distortion of the grain around it. Fortunately most timbers are not too difficult for most of us to attempt working with them.

Some species are a little soft or splintery and may need special care in planing, moulding or nailing and screwing. This point is noted where necessary in the timber descriptions.

MOISTURE, MOVEMENT AND DRYING

The previous section mentioned that wood is hygroscopic; this can cause trouble. The level of humidity in the air varies with country, season, weather conditions and whether indoors or outside. Wood will always try to attain an equilibrium with the conditions we put it in – technically this state is called the Equilibrium Moisture Content (EMC).

The moisture content of wood is always expressed as a percentage figure. It is a ratio in fact – the weight of water present relative to the weight of the dry timber. When a tree is felled the wood is very wet – there is more weight of water present than there is weight of wood; the moisture content may be anything up to 200 per cent or more.

Much of this moisture is described as free water and is present in the cell cavities. During drying this moisture level gradually drops and so of course does the weight. Eventually a level of about 30 per cent is reached; this level is called the Fibre Saturation Point (FSP) and below this moisture in the cell walls (known as bound water) is being removed. As bound water disappears, the cell wall contracts and the wood shrinks. How much it shrinks depends upon how low a moisture content is wanted and what the species is. The shrinkage is on average about 3.5 per cent in the radial direction and 7 per cent in the tangential direction; the difference between these figures can cause flat-sawn boards to curve across their width (cupping) while quarter-sawn boards get a bit smaller but remain flat.

Everything depends upon the humidity of the air because timber that has been well dried will become wetter on exposure to wetter air. This will allow swelling to occur as the timber approaches 30 per cent moisture content. Since the air humidity is constantly changing the EMC of wood is also changing; timber is therefore either swelling or shrinking whenever conditions change – this is known as movement. Some timbers are more stable than others and are better for changing conditions such as external joinery. The best timbers from this point of view may be too expensive for a particular job and so movement which opens joints and causes warping (distortion) and swelling is commonplace.

The strength of wood greatly increases as it is dried below the 30 per cent level. Published figures for strength values are normally quoted at 12 per cent moisture content. Suitable moisture contents for various uses in the United Kingdom are listed below. Failure to have wood at the correct moisture content to start with can only lead to trouble through movement.

20 %	Fungal damage unlikely below this level
18 %	Exterior joinery
14 %	Woodwork in buildings with little heating
12 %	Woodwork in buildings with regular intermittent heating
11 %	Woodwork in buildings with continuous heating
10 %	Woodwork in buildings with high level of heating
9 %	Woodwork close to heat sources or blocks laid on heated floors

Much antique furniture and old panelling have been damaged because central heating installations have dried wood to a lower moisture content and so caused shrinkage and often splitting.

When a log is first converted the wood is said to be green – undried. The cut boards can be dried out of doors with a roof on the stack to keep rain off. In the United Kingdom such drying is most unlikely to achieve a moisture content much lower than 20 per cent and it will take a year for each 25 mm of thickness. The alternative is to dry wood in a kiln where the humidity, airflow and warmth are controlled through a sequence of changes known as a drying schedule. Using a kiln, the process will only take a few weeks to even dryness and any desired moisture content can be obtained accurately and reliably.

DURABILITY AND PRESERVATION

The waste materials which accumulate in the heartwood and give wood its colour in many cases are poisonous to fungi and insects. The natural resistance to attack that a species possesses is known as the durability. Strictly the word should be used only for resistance to fungal attack.

Timbers are categorized according to how long their heartwood would resist attack when in contact with damp soil:

Very durable	– more than 25 years (Afrormosia, Teak, Iroko)
Durable	– 15 to 25 years (Western Red Cedar, European Oak, Agba)
Moderately durable	– 10 to 15 years (Pitch Pine, African Mahogany, Walnut)
Non-durable	– 5 to 10 years (most softwoods, Elm, Abura)
Perishable	– up to 5 years (Ash, Beech, Birch, Ramin)

The sapwood of almost all species is perishable, or non-durable at best. Preservative chemicals can be applied and so convert a perishable timber to a long-lasting one.

Fungal attack

Fungi are plants that cannot manufacture their own food so feed on other plants and animals instead. They require oxygen, a reasonable temperature and a moisture level of 20 per cent or over. Their reproductive spores float in the air and can land anywhere; if the conditions are right they will soon cause a lot of trouble.

Moulds are simple fungi that only grow on surfaces and do not penetrate wood. Damp storage conditions allow them to develop so they act as a warning.

Staining fungi attack sapwood that has not been dried and cause blue/grey discolouration. They only digest cell contents, as do moulds, but can ruin the appearance of timber. No real damage is done, but again they are a warning.

Soft rots are slow-growing specialized fungi that only penetrate about 50 mm from the surface; as the surface wears away they will penetrate more deeply. Soft rot is found on unprotected timber out of doors – fences, gate posts and exposed timber in buildings. It is also a severe pest of cooling towers.

Brown rots are a large group that are all able to digest the cellulose in wood but not the lignin. They cause wood to soften, crack along and across the grain to form cuboidal shapes and eventually to lose all its strength, but not its overall shape.

The most well-known brown rot is Dry Rot (*Serpula lacrymans*). This occurs in buildings and is very difficult to eradicate. The fungus produces a lot of fluffy white growth on surfaces when conditions are very damp; this fluffy mycelium often has drops of water like tears on it (hence the name *lacrymans* which means weeping). Under drier conditions there may be only small amounts of surface growth, possibly tinged with lilac or yellow. This fungus is particularly dangerous because it forms water-conducting strands which may travel a long distance over or through plaster and brick until it finds more timber. This timber, too, will be attacked, even if it is dry. Eventually the fungus will form a plate-like fruit body which produces brick red spores in enormous quantities.

Affected timber such as flooring and skirting becomes distorted and often smells musty. To get rid of Dry Rot requires very thorough removal of timber at least a metre beyond the visible signs, searching for strands in plaster and treating the whole area, including the foundation walls and soil, with a fungicide. It should go without saying that the cause of the original dampness must be put right, too.

The much less important Cellar Fungus (*Coniophora puteana*) is more common in buildings. There is rarely any surface growth and its strands cannot conduct water, but otherwise its effects are very similar to those of Dry Rot. The cure consists of putting right the cause of dampness, allowing everything to dry out and replacing affected timber where necessary.

White rots are also a large group. These digest all the components of wood and so destroy its shape and structure altogether. Before total removal the affected wood will be spongy and fibrous. It is often the cause of attack to external joinery such as window sills or door panels. Brown Rots are more likely to attack constructional softwoods.

All brown and white rots apart from Dry Rot are often given the general name of Wet Rot.

Insect attack

Keeping timber dry enough will always safeguard it from fungi but not from insects. There is always an insect which will like your wood, whether it is old or new, sapwood or heartwood, softwood or hardwood, temperate or tropical. All insects share the same life cycle; an egg is laid on the surface or in a crack; this hatches to produce a grub-like larva which bores into the wood. The length of the larval stage depends on the type of beetle but eventually the larva will move near the surface and undergo metamorphosis which converts it to an adult. The adult bites out an exit hole and on reaching the surface will mate and so produce more eggs. Adults can fly and so spread themselves easily. The damage to timber is usually caused by the larval stage but in some cases by the adults.

Power Post Beetle (Lyctus spp.) is found in the fresh sapwood of a few coarse-textured hardwoods such as Elm and Oak. Softwoods and hardwoods over a year old are immune. The wood is reduced to a fine powder like talcum and exit holes 1–2 mm in width are found. The damage can be severe in timber yards.

The Furniture Beetle (Anobium punctatum) is best known to most people as 'woodworm'. The adults lay eggs in the summer and the larvae tunnel in the wood for two or more years. They create a gritty dust and the emerging adults leave holes $1\frac{1}{2}$–2 mm wide. Polished surfaces are safe but attack can start at the back or on unfinished edges of a piece of wood. Any type of wood is attacked, including floors, structural and joinery timber.

The House Longhorn Beetle (*Hylotrupes bajulus*) is found only in certain parts of Surrey and on the European continent where it can be so serious that Building Regulations require preservative treatment for all new building work. The beetle is about 15 mm long, not counting its antennae; the larvae tunnel into softwood for up to eleven years making a coarse dust with long particles and the adults emerge through oval holes up to 10 mm wide. Few holes are made and the surface of the timber is left as a thin skin – often the first sign of infestation is collapse of part of a roof. The insect is very fussy about temperature at breeding times and the required conditions are fortunately not widespread.

Ambrosia Beetles (many different genera) are found in most countries, but are widespread in the tropics. The adult beetle bores the tunnels and grows a type of fungus on the cell walls. The fungus causes an obvious dark stain around the holes and is used as food for the larvae. Eggs are laid and larvae reared in the wood, quickly turning into adults. Adults vary in size and the damage is often called needle-hole, pinhole or shothole according to the size of holes found in the converted timber. Attack actually takes place in the freshly felled log and dies out as the timber is dried and the fungus dies. Wood showing the damage is known as 'sound wormy' and is perfectly safe to use provided its appearance is not important. It is sometimes found in the inside layers of plywoods.

Marine borers

Logs or timber which have floated in sea or brackish water may develop some damage from either Shipworm or Gribble. These two borers also attack piers, jetties and wooden boats.

The Shipworm starts as a free-swimming larva which bores into a piece of wood. Gradually it grows a worm-like body which may be a metre long and as thick as a finger. The hole is made larger to allow the body to grow and is often lined with a chalky deposit. The damage is out of sight and when severe enough causes sudden collapse.

The Gribble is like a small shrimp. It burrows into wood near the water level and wave action removes the weakened surface. As the gribble bores in more deeply each time the wood gradually assumes a shape like an hour-glass and the damage becomes obvious.

A few species of timber are resistant to borers but it is often best to use a good creosote treatment.

Termites

Termites, otherwise known as white ants, are common in tropical and subtropical areas but are also found in temperate regions. So far they are not known in the United Kingdom but they do exist in France and Germany. There are many species known, which fall into two distinct categories.

Subterranean termites are the most widespread type and are responsible for most of the damage. They nest out of doors and travel in mud tunnels which they build over surfaces to reach wood. Houses need careful design to prevent this tunnel building or else the surrounding soil needs to be poisoned.

Drywood termites never enter the ground. They live in dry wood only; the infestation is started by winged forms laying eggs in cracks and joints of structural timber and furniture. An intact skin is left on the surface, so again attack can be severe but not noticed until collapse occurs. Only preservation techniques will protect wood; some species of timber are naturally resistant but the resistance varies in different countries, according to the species of termite that lives there.

Preservation

In many situations and for a variety of reasons it is safer not to rely upon careful maintenance of timber or natural durability but to apply a chemical preservative either before manufacture or after.

It is simple and cheap to apply solutions by brush or spray but the treatment will not be very deep unless the wood is very permeable – most woods are not, except for their sapwood. It is better to treat the wood by a pressure process after it has been dried. The pressure will push the solution further in than any non-pressurized process can achieve; if the wood is put in a treatment cylinder and a vacuum is applied before running the solution in and applying pressure, then an even better penetration will result because there is no internal air pressure to resist it. Applying a second vacuum phase after draining the preservative will take out surplus solution for re-use. There is, after all, no point in filling cells with solution – it is only the walls which need protection. This double vacuum process is widely available and it is easy to buy treated timber. It is better to do all possible machining before treatment, otherwise all cut surfaces should be treated in case an untreated core has been exposed.

An ideal preservative would be: effective against all pests, permanent, stable, cheap, easy to obtain and apply, deeply penetrating, safe, free from smell or tainting or staining, not affecting wood properties or machining, not needing re-drying, able to be glued or painted. We can choose from the following types:

- Creosote – good for outdoor use but smell and staining prevent its use inside.
- Water-borne solutions of copper, chromium and arsenic chemicals. These fix well to the wood and are usually pale green in colour. The timber has to be re-dried after pressure treatment and may corrode metal fixings in time. Not expensive. One treatment uses boron solution which is a safe chemical but it takes several weeks for the chemical to penetrate well and then be dried so it is less popular.
- organic solvent solutions are mixtures of tributyl tin oxide or pentachlorophenol (fungicides) and Dieldrin or Lindane (insecticides) dissolved most often in white spirit. There is a slight fire risk while the solvent dries off but the chief problem is the higher cost of the chemicals and solvent. Timber can be brought already pressure-treated or the solution can be used for brush or spray treatment of existing structures.
- The chemicals used in organic solvents can also be used in a water-based emulsion form. This is widely used by remedial treatment companies; it is spread thickly on all surfaces of constructional timbers and simply soaks in over a period of months. This treatment is not suitable for visible timbers.
- Appropriate chemicals can also be applied to give flame-proof protection and water-repellancy to timber.

STAINING AND FINISHING

A very wide variety of finishes is available both for the professional and amateur markets and good results are easily obtainable.

In general, it is a matter of sealing the surface so that dirt, wear and weathering can never affect the prepared surface. It is usually necessary to renew the coating from time to time, especially for exposed timber.

Stains can be applied in water or spirit. Water-based colours tend to dry patchily and often roughen the surface; the need to smooth the re-dried surface often further disturbs the colour. Spirit-based stains penetrate well and dry rapidly with no swelling effect.

Clear finishes can be applied over stains or to the natural surface. The choice depends upon conditions of exposure and wear. Stained finishes are also available but have the disadvantage that if they become chipped or scratched the natural colour of the wood will show through.

Paints are available in great variety. Here the whole point is to obscure the wood and the only likely trouble is that the few gum- or resin-containing species might cause some discolouration in time. Knots are a risk in this way and should be sealed first with special knotting primer. Many woods are too permeable to apply undercoat and topcoat and get a good finish; a primer is needed to seal the wood first because the cellular structure sucks the paint in. The recently developed microporous paints allow moisture in the wood to dry out through the paint without causing blisters; unfortunately they also allow moisture from rain to enter the wood through the paint film and so maintain a dangerously high moisture content from the point of view of possible fungal attack for several months of the year.

Water repellants, usually based on silicones, are a very simple finish to apply to exposed joinery and give good protection. They can be bought in a range of colours, or clear.

STRENGTH AND QUALITY

Wood is regarded as being very strong for its weight and it is used for a wide variety of purposes. Roughly speaking, the strength is proportional to the weight, so weight is the single best guide to likely performance – timbers of similar weights are likely to give similar performances, provided the same sized sections are used.

For any given species it is possible to calculate the strength that a particular size section will have on average and these are the values that should be used for any type of construction. It is no use for a designer to say he would like to reduce the size of a chair leg for instance – the smaller size might carry a certain load but the legs might not be stiff enough if the chair was tipped on two legs only.

Very frequently larger timbers are used than is strictly needed for many small jobs, partly because it does not look right if a piece has too small a bulk. For building and other constructional purposes sizes that were unnecessarily large were often used in the past. The beams in a cottage show that clearly. In more recent times smaller sections have been used in order to save money on the cost of materials; it is necessary though, not to take this too far or frameworks will begin to sag and creep when under load.

Apart from the species selected for any purpose, care has to be taken over the quality of each piece; it is always possible for defects to be present, or some abnormality of growth due to adverse soil or weather conditions at some stage during the growth of the original tree. Provided that the weight is normal for the species, not too much will be wrong but other factors need to be considered too.

- Grain. The direction of grain in a piece is very important; sloping grain will reduce strength in compression or in tension and therefore in bending too. Small amounts of slope or local irregularities will not matter unless it is a very small section of timber.
- Knots. These are the bases of branches which have been grown over by the tree trunk as it gets fatter year by year. Each knot is both a 'missing' piece of the normal grain and causes a deflection of the grain that is present. The number and size of knots and their position in a piece are also vitally important.
- Moisture content. This is also important in that strength develops progressively below 30 per cent. If a dried piece of wood that is only just strong enough to carry its load should become wetted then it may well fail. It is usually safer to over-design in terms of size in case constructional timbers do get wet through flood or storm damage. In all fairness it should be pointed out that modern buildings do not have dangerously small timbers because Building Regulations would not permit it.

Grading

It must be recognized that even a single species may be so variable that it is necessary to put different pieces of it into different quality groups, or grades. The buyer can then select the most suitable and affordable grade and choose from the range of common sizes available as converted stock.

Softwoods are graded into six qualities by visual inspection at the sawmill; the inspector sees all the surfaces and estimates the effect of knots and other defects such as splits, distortion, sawing defects, presence of fungal stain or rot and moisture content. The graded pieces are then sorted and dried and then graded again. Each piece then has a shipper's mark printed on its end; this mark indicates the source of the wood and usually its grade too.

Each country of origin has its own grading rules and they are not all identical. Very litte mature virgin forest now exists to give large supplies of the best softwoods so it has become normal practice to sell Grades 1, 2, 3 and 4 from Scandinavian countries as Unsorted grade – the four grades are mixed in the proportions of 5:10:65:20. The equivalent material from Russia consists only of Grades 1,2 and 3. Canadian systems are different again and even depend upon whether the stock comes from east or west coasts. The situation is complicated even for those who know it well and so much reliance has to be placed upon the actual supplier to provide material fit for a given purpose.

The systems outlined above are known as commercial grading and are based upon the appearance of a piece from the point of view of being used in construction. Hardwoods are graded on the same principle but from the point of view of their decorative value. Inspectors examine the upper and lower surfaces (faces) of each board and ignore the edges; they decide basically how much clear, usable stock there is in each piece. Hardwoods do not have so many defects as softwoods do and knots are rare. There are many more species available, from many more countries and they are cut into a much wider range of sizes, so the position is rather more complicated than for softwoods. Hardwood grades are also based upon the size of each piece as well as quality, so effectively the system is deciding upon the percentage surface area which is free of any defect, together with the length and width allowed for the particular species.

The standard hardwood grades are: Firsts, Seconds (usually sold as a combined grade – F. A. S.), Selects, No. 1 Common (usually sold together as No. 1 C & S), No. 2 Common, Sound Wormy (safe to use but blemished), No. 3A Common and No. 3B Common. Remembering that each of these grades have different lengths and widths and all countries are slightly different, it is easy to see that the situation is complicated and the supplier's advice is often necessary.

The increasing use of timber as a structural material has led to the development of stress grading of softwoods as distinct from appearance grading methods. This is the prediction of its strength in a more precise and reliable way. Trained graders estimate the effect of knots and other defects in a three-dimensional way; timber is categorized in this way into General Structural (GS) or Special Structural (SS), or it is rejected. It takes longer to grade in this way so the cost is higher.

A more accurate system is available using a machine to load each piece of timber separately and so actually measure its load-bearing capacity. This produces MGS and MSS grades at a higher price still.

SHEET MATERIALS

A wide variety of wood-based sheet materials is available. They have been developed because they provide characteristics of size, stability and strength that are not so easily available in solid timber. In some cases they are able to be manufactured from material that is otherwise useless, so additionally they prevent waste. Considering the area they cover with little effort they are not expensive and offer many advantages.

There are three main groups:

Particle boards are made mostly from wood chips mixed with adhesives. According to the size and shape of the chips and the way they are arranged in the sheet, a wide range of properties results. The lowest grade of chipboard, as they are commonly called, is suitable for veneering as furniture but is too weak for heavy loading as shelves. Relatively cheap, it needs careful choice of type for each job.

Fibreboards are made from wood that has been mechanically ground into little clumps of fibres. The wet fibre is laid into a mat, cut to length and each sheet compressed in a hot press. The heat dries off the water and the fibres bond themselves to each other without the aid of adhesive. Slight pressure produces softboard about 25 mm thick, used for ceilings and insulation. Medium pressure gives a board about 10 mm thick which is ideal for noticeboards and partitioning. Heavy pressure provides hardboard which is 2–4 mm thick; it is used extensively for doors, caravan linings and shop fittings.

A fairly new type of board is now very popular; it is made from fibres and an adhesive and is known as medium density fibreboard – MDF. This material is made in a wide range of thicknesses and is virtually wood with no grain or decorative appearance. It can be cut, machined, glued, veneered, stained and painted just like solid wood and is now widely used for furniture and some types of joinery.

Plywoods are made from thin sheets of wood veneer cut by literally unrolling a log – it is turned against a large fixed knife to produce an endless sheet which is cut into lengths, dried, glued and assembled into layers with the grain at right-angles in successive sheets. Most sheets have a balanced construction with an odd number of layers.

A seemingly endless variety of thicknesses, numbers of layers and range of single or mixed species exists, not to mention the types of glue that can be used. Sheets will broadly be categorized into weather and boil proof (WBP) suitable for outdoor use; boil resistant (BR) suitable for sheltered outdoor use; moisture and moderately weather resistant (MR) and interior (INT) suitable only for short life outdoors but long life indoors. Any plywood can be faced with a special quality or figured decorative veneer, as can most chipboards or MDF.

Some special boards called core plywoods exist. These have battens or strips of solid wood inside the surface veneers. They tend to be stiffer, thicker and more expensive. They make good partitions, small doors (when the edges are covered), or they can be used in good quality furniture.

The advice of a supplier is again invaluable when choosing sheet materials. Strength and appearance do not necessarily go together. Some sheet materials have been given a bad name when really the wrong type was used for a particular job – a false economy, very often. For instance it is undesirable to stick tiles for a shower cubicle onto chipboard – the chips will swell and the board will fail. It is much safer to use a WBP plywood and this will prove to be cheaper in the end. Again, shelves need the best grade of chipboard or a core plywood – other types will slowly droop under the load.

DESIGN AND UTILIZATION

For many purposes wood is used simply because it looks good and its other properties are of such little importance that they are not taken into account. In these cases design is a matter of pleasing or satisfactory appearance and probably has no structural implications at all.

Where timber is used for buildings, large furniture, vehicles, containers or sports goods then irrespective of whether there is an appearance requirement there is certainly a strength or performance need. To a large extent this is taken care of in grading systems but often it is necessary to calculate precise loadings in order to determine the size of a particular component in a structure. Guesswork will not do and so architects and timber engineers need to be familiar

38

not only with published strength figures but also the additional factors that arise because wood is a variable biological material.

For certain purposes a factor such as durability may be of prime importance and so must be considered first by selecting from timbers in the correct category, or by accepting that a preservative must be applied.

A great deal of published information exists and advice is available from a number of research and other organizations. Some useful addresses are listed at the end of this chapter.

There is a tendency to use a limited number or even a single species of timber for a given purpose. With shortages of many species developing through over-use manufacturers will certainly have to accept that a wider range of species is not only necessary but is perfectly safe and satisfactory. In the timber descriptions which follow the principal uses are listed; many of these uses are those for which the timber is used in its country of origin, although it may have a more restricted use elsewhere. If cost or availability are ever a problem it is not at all difficult to select an alternative species that will perform just as well, even if it looks rather different – and appearance can always be altered if necessary.

A separate, but closely involved aspect of utilization, is choice of joining technique. Traditional mechanical machined joints have often given way to the use of one of a vast range of metal or plastic connectors that may incorporate nails, screws or bolts. Adhesives of great strength also exist. Utilization is a matter of correct selection here too – it is useless to have a strong, durable timber if fasteners or adhesives are not equally durable on exposure to moisture or weathering.

USEFUL REFERENCE ADDRESSES

BUILDING RESEARCH
ESTABLISHMENT
Building Research Station
Garston
Watford
Herts WD2 7JR

BRITISH WOOD
PRESERVING
ASSOCIATION
Premier House
150 Southampton Row
London WC1B 5AL

FINNISH PLYWOOD
INTERNATIONAL
PO Box 99
Welwyn Garden City
Herts AL7 0HS

FURNITURE INDUSTRY
RESEARCH ASSOCIATION
Maxwell Road
Stevenage
Herts SG1 2EW

TIMBER RESEARCH AND
DEVELOPMENT
ASSOCIATION
Stocking Lane
Hughenden Valley
High Wycombe
Bucks HP14 4ND

THE TIMBER TRADE
FEDERATION
Clareville House
26/27 Oxendon Street
London SW1Y 4EL

Part Two

DESCRIPTION OF COMMERCIAL SPECIES

3 Hardwoods

Many of the entries have small drawings of end-grain structure such as the ones below. In all cases rays are shown running up and down the picture and growth rings go across. Generally the width of a single growth ring is shown.

This shows a ring-porous hardwood with tyloses in some vessels. Storage tissue (parenchyma) is shown in the dotted areas.

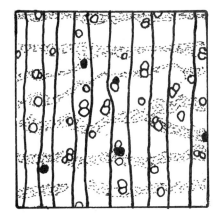

This is a diffuse-porous hardwood. Parenchyma is present as distinct, slightly wavy bands which include many of the more or less evenly distributed vessels. Growth rings may or may not be countable or obvious, depending on the species. Gum deposits are present in some of the vessels.

Throughout this chapter the parenchyma is shown as areas or lines of small dots. Tyloses are shown as a cross in the vessel. Deposits of gum or minerals are shown as filled-in vessels. For many species more than one name is available. Wherever possible an entry is placed alphabetically under its Standard Trade Name; other names are listed if they are likely to have been used in the United Kingdom.

Weights are given in metric terms and are quoted for 12 per cent moisture content. If the figure is required in imperial terms (lb/foot^3) then it should be divided by a conversion factor of 16.

Durability comments relate to heartwood exposed to fungal decay. Sapwood is always perishable. (Details of durability ratings are given on page 28.)

Insect attack can always happen (unless the wood is treated) and will affect both heartwood and sapwood.

Treatability comments again refer to heartwood. Sapwood is usually easy to treat in most species.

Movement is any shrinkage or swelling that occurs after the initial drying, while the timber is in use. The actual percentage values of the size changes possible are mostly small (in the region of 1, 2 or 3 per cent for small, medium and large movement) but they may matter a great deal in the maintenance of a good appearance over a period of time.

Abura

(Bahia, Subaha)

Mitragyna ciliata

This timber is a rain forest species from West Africa. It is a uniform light brown colour, often with a yellowish or pinkish tinge; the sapwood is not obviously different and seems to form most of the tree. The grain is usually straight but may be interlocked and therefore likely to pick up in planing. The weight is variable, averaging about 560 kg/m^3 when dry.

Movement	–	small
Durability	–	perishable
Treatability	–	moderately resistant
Workability	–	generally good but sometimes abrasive on cutting edges

A useful plain timber that behaves well. Popular for non-showwood in furniture, it can also be finished easily to match almost any veneer. Also used for interior joinery, mouldings and some plywood.

The wood has a strong, unpleasant smell when still wet; this disappears when drying is complete. A few people have found the dust produced during working a little irritant to the skin but this soon disappears. May have brittleheart.

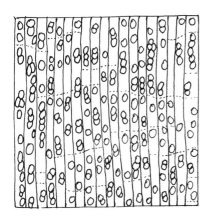

Structure is very plain and simple. Sometimes fine lines of parenchyma connecting the rays can be seen.

Acacia
Various *Acacia* species

This genus is very widely distributed and the timbers of many different species are to be found in commercial consignments of the wood, though specific varieties may be sold under special names (for example *Acacia melanoxylon* reaches the market as Australian Blackwood). It is a coarse-textured, reddish-brown wood, sometimes showing a figuring of dark streaks. The species tend to be hard and heavy.

Movement   –   variable
Durability   –   good
Treatability   –   not known
Workability   –   difficult because of its hardness but it finishes well

It is a strong and elastic timber that may be used for the making of tool handles, vehicle parts, walking sticks and turned articles.

Afara   (White Afara, Ofram, Limba)
Terminalia superba

This West African timber is sometimes known as White Afara to differentiate it from a close botanical relative Idigbo (or Black Afara), described later. Both the names Black and White Afara are best not used to avoid confusion. Two distinct varieties of this timber are available. In one the heartwood is of a uniform pale yellow or brown; in the other the timber is strongly marked with irregular olive or greyish streaks: the names Limba Clair and Limba Noir respectively are used for timber from French Colonial sources to distinguish between the two. Afara has an irregular grain, with a moderately coarse but even texture. Its weight is rather variable but is normally between 480 and 640 kg/m^3 when dried.

Movement   –   small
Durability   –   non-durable
Treatability –   moderately resistant, needs pressure
Workability –   easy to work but grain may pick up; finishes well

Widely used for its decorative value, a lot of Afara is made into plywood. Also used for furniture, interior joinery and fittings whether plain or streaked. Pinhole borer or brittleheart may be present in logs or timber.

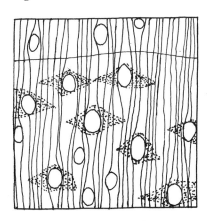

Vessels are large enough to need filling before polishing. Terminal and aliform parenchyma are very obvious.

Afrormosia (Kokrodua)
Pericopsis elata

The sapwood and heartwood of this West African timber are not easily distinguishable. The sapwood is very narrow and of a slightly paler colour than the heartwood which is brownish-yellow (sometimes with darker streaks) after being exposed to the air for some time, though it is plain brown in colour when first sawn. The grain is variable from straight to interlocked, while the texture is close and even. The average weight for seasoned stock is about 690 kg/m³ when dried.

Movement   –   small
Durability   –   very durable and also insect resistant
Treatability   –   very resistant
Workability   –   works well except for pick-up with interlocked grain; splits with nailing; polishes well

Afrormosia is used for high quality interior or exterior joinery and fittings, decorative block flooring, furniture, etc. and has also been used as an alternative to Teak for certain purposes. It is a strong wood that seasons without difficulty. It is generally thought to look like Teak but is really a little darker and finer textured. The slow darkening can be annoying when matching new timber with older pieces. Available as plywood.

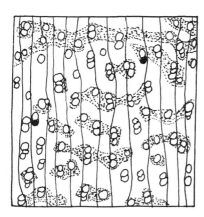

Vessels are very hard to see without a lens. Ripple marks show slightly on flat-sawn surfaces.

Afzelia

(Apa, Doussié, Chamfuta)

Various *Afzelia* species

Trees producing the timber sold under this name are widespread throughout Africa. Commercial consignments may be made up of *Afzelia africana*, *Afzelia quanzensis*, and other *Afzelia* species. It is a 'general purpose' timber, suitable for structural work and flooring as well as for good-quality interior joinery.

Afzelia is a coarse-textured wood of rather plain appearance, red-brown in colour and with a grain that may be straight to shallowly interlocked. Density is somewhat variable because of the range of species but averages at 820 kg/m³ when dried.

Movement	–	small
Durability	–	very durable but slow to dry
Treatability	–	very resistant
Workability	–	sometimes difficult because it is hard; finishes well with care

Afzelia contains a yellowish dye which will cause staining if other materials are brought into contact with the damp wood. Used for heavy constructional work (especially docks and bridges), high class joinery, vehicle framing and flooring.

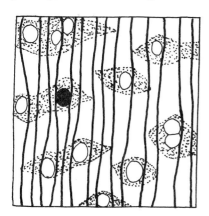

Yellow or white deposits may be present in vessels. Parenchyma is aliform and very obvious.

Agba

(Tola, Tola Branca)

Gossweilerodendron balsamiferum

The sapwood and the heartwood of this species are not well defined, the timber being of a rather featureless, yellowish-pink to reddish-brown shade. Planed surfaces are mildly silky but may be marred by copious exudations of gum unless it has been carefully selected. The grain is straight, the texture fine and even and the average weight is about 510 kg/m³ when dried.

Movement   –   small
Durability   –   durable but exposure to sun may bring out gum
Treatability   –   resistant
Workability   –   easy to work unless much gum is present; stains and finishes well

This wood is very like a fine-textured Mahogany and is often used for furniture. Also suitable for internal and external joinery, flooring and plywood. A very useful general purpose timber, that is even safe for boat building. Brittleheart may be present.

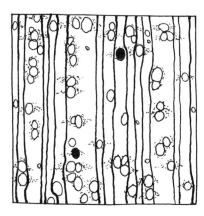

Gum deposits are present in some vessels. The occasional holes that look like small vessels are actually gum canals.

Albizia, West African

Various *Albizia* species

(Okuro, Ayinre)

This timber of tropical Africa has taken its English trade name from its botanical identity, as it is the product of at least three *Albizia* species, *Albizia adianthifolia*, *Albizia ferruginea* and *Albizia zygia*. The sapwood is yellow in colour and the heartwood varies from mid-brown to dark chocolate brown. The grain may be either interlocked or irregular, with the texture moderately coarse. The weight ranges from about 580 to 820 kg/m³; the average is about 700 kg/m³ when dried.

Movement   –   small
Durability   –   very durable
Treatability   –   very resistant
Workability   –   fairly easy except with severely interlocking grain; needs pre-boring.

Suitable uses would be joinery and general construction work.

This wood produces a very fine sawdust which may cause irritant effects to the nose unless a mask is worn.

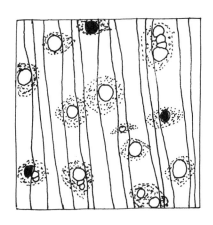

Deposits are often present in the vessels.

Alder (Black Alder, Grey Alder)
Alnus glutinosa

Differing species of *Alnus* go to make up consignments of this timber but they seem to be indistinguishable. The tree is common in Britain and the northern parts of Asia and America and it produces a straight-grained, fine textured wood. The timber may be marketed under the description of 'Grey', 'White' or 'Red Alder' according to species but in general the wood ranges from light to reddish-brown in colour, with a dull surface and weighs about 530 kg/m³ when dried.

Movement   –   probably reasonable
Durability   –   perishable
Treatability   –   good
Workability   –   easy to work and takes stains and finishes well; turns well

Alder may be used for such purposes as cabinet and furniture making, plywood manufacture, shoe heels, clogs, bobbins, wooden cogs and small turned items. Sap-stain fungal infection is not unusual with the species and evidence of its attack may be present in the converted wood. Small brown pith-flecks are quite common on surfaces.

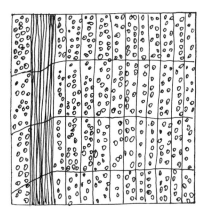

Growth rings may have a wavy outline. Rays are unusual in that they are sometimes grouped very closely into aggregates which give the appearance of being single very large rays. Scalariform perforations are present.

Alder, Western Red
Alnus rubra

This timber is often described merely as Alder but is more likely to be marketed under the more precise name given above. This is the most important of the North American *Alnus* species and in the USA it may be sold as Oregon or Red Alder. It was formerly the custom to bleach the timber and then sell it under the name of Basswood, leading to confusion with *Tilia glabra* and other species given the same name. An average specimen of Western Red Alder weighs rather more than does one of common Alder and in colour the American timber may range from a pale yellow to reddish-brown, with an indistinct figuring.

The tree has its natural habitat on the Pacific coast and the timber is a useful non-ornament cabinet wood that is largely used for furniture making and turning. It is also extensively used for the manufacture of plywood. Western Red Alder is not a naturally durable timber when exposed to the attack of wood-rotting fungi but is durable for use under water. It works satisfactorily in all hand or machine operations and finishes well.

With a hand lens the wood is not distinguishable from *Alnus glutinosa*. Scalariform perforations are present.

Almond
Prunus amygdalus

Commercially, this species is far more important for its fruit than for its timber. Reddish in colour and with a light, lustrous surface, the grain may be either straight or irregular but the texture is fine and uniform. It is not naturally durable for use in exposed positions but is not normally given a wood preservative treatment. In most hand and machine operations Almond cannot be classified as an easy timber to work but it may be brought to a good surface and also carves and turns well.

The wood dries readily and is suitable for inlaying, small turned articles and fine cabinet work. The tree is common in Europe and East Africa but even under the most favourable conditions it only grows to a height of some 10 metres and is more commonly much less, so the wood is available only in small sizes. Apple and Cherry are two species that are closely allied to the Almond; the three timbers have much in common but Almond is the least commonly seen.

Aloes Wood
Acquilaria agallocha

This species has been known under such exotic names as Paradise Wood, Eagle Wood or Calambac. The timber comes from a tall tree that is native to China, Malaysia and other parts of tropical Asia. Nowadays the wood is not well known on the world markets, though once it was in great demand for inlaying and similar work. Aloes Wood is an easy enough one to work but needs pre-boring before nailing or screwing as it has a very marked tendency to split. Under drying treatments it may check badly and it is also subject to chemical stain during drying. The sapwood of the species is whitish in colour and has no commercial importance but the heartwood is dark coloured, with a beautiful and distinctive figuring and a fragrant, resinous smell. Average dried specimens weigh about 400 kg/m^3 and the grain is straight but the texture medium coarse.

Alstonia   (Patternwood, Stoolwood, Duku, Sindru)
Alstonia congensis

This timber is best known from West Africa but occurs right across to East Africa. It is a soft and light timber weighing about 400 kg/m³ when dried. The sapwood and heartwood are both almost white but become yellowish on exposure. The grain is straight and the texture is fine and even. May show signs of fungal staining which occurs before drying.

Movement   –   small
Durability   –   perishable
Treatability   –   permeable
Workability   –   very easy to work provided all cutting tools are sharp

This wood shows an unusual structural feature. It is the regular presence of latex traces which show as slits resembling knife marks on the flat-sawn surfaces; often dried latex is still present. These slits are very unsightly and so the timber is normally used for rough work or for the core strips of blockboard.

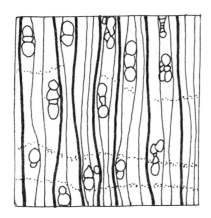

The latex traces will only show on an end-grain if they happen to have been cut along their length.

Amboyna
Pterocarpus indicus

(Narra, Angsama)

Amboyna is a timber of attractive appearance that is most commonly used in veneer form, these being obtained from burrs; a small percentage of solid stock is available and this may be used for high-class fittings and furniture. The timber is closely related botanically to the padauks. The description given below applies to the solid wood.

The clearly demarcated sapwood is whitish when freshly cut and though darkening a little on exposure it remains clearly defined from the heartwood, which is variable in colour from pale yellow to blood red and is normally very highly figured. Grain is either straight or interlocked and the texture is moderately fine and even; the planed timber has a mild, attractive smell and a slightly silky surface. The weight may be rather variable but averages 560 kg/m³ when dried. In its working properties the wood behaves very similarly to the well-known Andaman Padauk; it may need some care in planing but can be brought to a good surface that will take stain well and can be given an excellent polish. It dries well (though the solid material may develop minute surface checks) and moves little when dry. A slight amount of wood-borer beetle damage may occur in the species, though the timber is strongly resistant to fungi and other wood-destroying agents. This is one of the most important export species of the Philippines (the wood being known as Narra on the North American market) and it also occurs in Malaya and North Borneo, where the local name of Angsama is used.

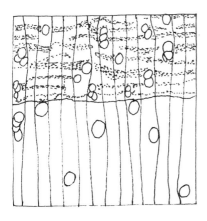

The chief difference between this species and Andaman Padauk is that the vessels of Amboyna are both smaller and fewer.

Angelin   (Partridge or Pheasant Wood)
Andira inermis

The other common names arise from a similarity in appearance to the barred colouring on the birds' feathers. The barred appearance develops as a result of the regular presence of terminal and banded parenchyma, alternating with bands of darker fibres. This is a species that is often highly decorative; the timber comes from the West Indies and neighbouring regions. This tree is of medium growth only and the timber is not always available in good sizes.

Angelin is a straight grained, moderately open-textured wood with an average dried weight of 800 kg/m^3. It is reddish-brown to almost black in colour, sometimes handsomely figured, though with the lighter, unfigured samples the sapwood may be difficult to differentiate from the heartwood. The timber is rather hard to work but can be brought to a good surface and will turn well; the response to stain and polish treatments is very good. It is a rather brittle wood. The timber is strongly resistant to fungal and insect attack and dries well. Ripple marks are sometimes visible. It is primarily a structural wood, though figured stock may be used for turnery, cabinet work and so on.

Aningeria   (Mukangu)
Aningeria species

Four species occur across Tropical Africa and may be marketed
separately or mixed, according to their origin; all are very similar.
The sapwood is not obvious; the wood is pale yellowish to light brown,
the grain mostly straight and the texture is fine to moderately coarse.
The weight is a little variable but averages about 530 kg/m³ when
dried.

Movement   –   no information
Durability   –   perishable
Treatability –   permeable
Workability   –   depending upon the amount of silica present, logs
                  may be fairly easy or difficult

The wood peels and slices well and is used for veneers if wavy grain
is present, or for plywood otherwise. Suitable for joinery or interior
construction.

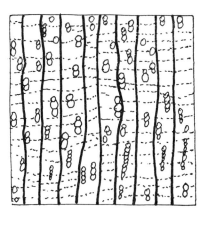

Vessels arranged in radial
multiples are often quite
prominent. Fine lines of
parenchyma together with the
rays form a net-like effect.

Antiaris

(Oro, Chenchen, Ako)

Antiaris toxicaria

This is another example of a timber using its botanical name as its common name. The species grows throughout tropical Africa. It is usually a rather low-grade timber, similar to Obeche but generally thought of as inferior. The sapwood and the heartwood are poorly defined; the sapwood is very wide and the timber as a whole is whitish to yellowish-grey in colour. Undried stock has a rather objectionable smell, the grain is interlocked and the texture coarse and rather uneven. Antiaris is a light-weight species, averaging about 430 kg/m^3 when dried.

Movement   – small
Durability   – perishable
Treatability   – permeable
Workability   – easy to work but its softness requires very sharp cutting edges; finishes well

This timber needs rapid drying or it easily develops fungal staining and insect attack. It can be used for light furniture construction, packing cases and is suitable for core plywoods. Quartered veneers show a good stripe figure.

Apple
Malus sylvestris

Like the Almond, the Apple tree is small and is economically more important for its fruits than for its timber, the more so as the tree may live and bear fruit for more than 150 years. The tree is a very common one but reaches its best development in the colder parts of the temperate zones. Although not greatly used for external work, the timber is naturally resistant to the attack of wood-rotting fungi. It is rated as hard and is somewhat difficult to work by hand but is an ideal timber for carving and it also turns well.

Apple responds very satisfactorily to the usual types of finishing treatment given to a non-ornamental cabinet wood and in particular polishes well. In general appearance the wood closely resembles that of Pear, having brownish-pink or light reddish coloured heartwood. The grain is normally irregular (sometimes sufficiently so to cause difficulty in machine planing) but the texture is fine and even. Average specimens weigh about 700 kg/m³ when dried. It is sometimes found in old furniture sold under the name of Fruitwood.

Ash
Fraxinus excelsior

(English, European Ash, etc., Olive Ash)

The Ash tree is one of the tallest growing hardwoods indigenous to Europe and its widespread distribution has led to such geographical prefixes as English, Swedish, Spanish, Polish, etc. Samples that have irregular dark streaks in the heartwood are often sold as Olive Ash; this effect is believed to be due to a non-damaging type of virus or bacterial attack and it looks very similar to the natural colouring of Olive. Normally the wood ranges from white to light brown with an indistinct sapwood and shows a strong figure due to prominent growth ring outlines. It is more variable in weight than most timbers; it ranges between 510 and 830 kg/m³ and averages 690 kg/m³ when dried. The grain is normally straight and the texture rather coarse. Ash is a particularly tough and elastic timber provided that it is of a normal growth rate (about 6–12 growth rings/25mm radius).

Movement   –   medium
Durability   –   perishable
Treatability   –   moderately resistant
Workability   –   works moderately well except in harder samples; finishes well; requires pre-boring

The wood steam-bends very well and turns satisfactorily for a timber of its unevenness of texture. Typical uses include the making of ladders, hammer and tool handles, spokes, oars, poles, sports goods, gymnasium equipment, coffins, garden furniture, vehicles, artificial limbs and other purposes for which a resilient but reasonably light timber is needed. Its bending properties make it useful in the furniture and boat-building industries but its perishability requires treatment.

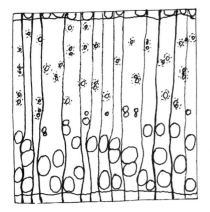

A ring-porous wood. Latewood vessels are single or in pairs, otherwise it resembles Sweet Chestnut.

Ash, White

Fraxinus americana principally

There is little essential difference between the structure and nature of this North American timber and European Ash, except the often light brown to reddish-brown timber has a rather more pronounced colour difference between the earlywood and the latewood and it may be a little lighter in weight. Working and seasoning qualities of White Ash are approximately the same. It is used for the making of shop fittings, gymnasium equipment, athletic goods of all descriptions, spokes and other parts of vehicles, tool and hammer handles and all the purposes for which European Ash may be used.

The wood sometimes reaches the market under the alternative descriptions of American White Ash or Canadian White Ash and American shippers sometimes differentiate consignments as Soft Ash or Tough Ash. In all, some ten or more different species may be sold under the blanket description of White Ash. *Fraxinus nigra* is usually marked as either Black Ash or Brown Ash (not to be confused with the decorative Olive Ash).

Avodiré
Turraeanthus africanus

A timber from West Africa which has been found suitable for high-class interior joinery. Selected stock is commonly used in the manufacture of decorative veneers. Sapwood and heartwood are only poorly defined, the timber maturing to a golden yellow colour which, in unfigured stock, gives it something of the appearance of a pale-coloured mahogany. The grain is sometimes straight but is often wavy or irregularly interlocked and this produces an attractive mottled figure on quartered material. The average weight is about 550 kg/m^3 when dried. The texture is medium.

Movement   –   small
Durability   –   non-durable
Treatability –   very resistant
Workability   –   easy to work except when grain disturbance is severe; finishes well; needs pre-boring.

The tree is a small one and so larger pieces of timber are not available. It is used for joinery and furniture in the solid and is very popular as figured veneers.

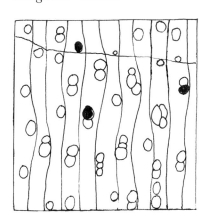

Growth ring outlines are rarely visible and no parenchyma can be seen. Gum deposits may be present.

65

Ayan

(Movingui)

Distemonanthus benthamianus

This wood has been sold as Nigerian Satinwood but since it is not a true satinwood the name should not be used.

It is an interlocked grained timber, so when cut on the quarter it shows a ribbon stripe figure. The heartwood is usually of a pale golden yellow to yellowish-brown colour, though sometimes irregular, slightly darker streaks are present. It is a lustrous, fine and even textured wood, having an average weight of about 670 kg/m³ when dried.

Movement   – small
Durability   – moderately durable
Treatability – resistant
Workability – not difficult but can dull cutting edges quite quickly; needs pre-boring; finishes well

This wood contains a yellow dye which may stain other materials under damp conditions. A good joinery and cabinet timber. Has been used for vehicles and flooring.

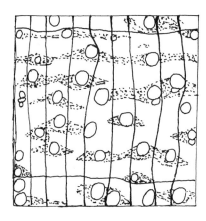

Vessel deposits sometimes present. Ripple marks usually easily visible on flat-sawn surfaces.

Balau (various)

Shorea species

This timber is also known as Sal, Chan and Selangan Batu according to the country of origin – details are given under the entry for Lauan. Red and yellow types exist.

The species in this group have an average weight of between 790–960 kg/m³ when dried. Their colour is either a yellowish-brown or a dark red-brown according to species and their texture varies from moderately fine to coarse – it is always even. The grain is always interlocked and the woods have strength properties similar to those of Greenheart.

Balau-type timbers are used for heavy constructional work, wharf construction, flooring, sleepers and boat framing. Red Balau is less durable than the others so some care is needed in selection for outdoor uses.

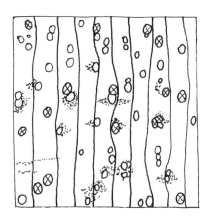

Sparkling tyloses in the vessels are common. A small amount of parenchyma of the aliform type is usually present.

Balsa   (Guano, Polak)
Ochroma pyramidale

This hardwood timber is remarkable for being the lightest weight species to achieve commercial use. It is very variable in weight, varying from 40 to over 320 kg/m³ when dried; commercial material is usually selected to be within the range 100 to 225 kg/m³. The tree grows widely throughout tropical America.

The wood is straight grained and coarse textured, being white, pinkish-white or pale brownish-white in colour and having a slightly lustrous surface. Being soft the edges of the wood have a tendency to crumble under dull tools and the same weakness makes nail and screw holding properties poor. Glue adheres well however and the wood may also be stained and polished with satisfactory results.

Movement   –   small
Durability   –   perishable
Treatability –   resistant
Workability   –   needs very sharp cutting edges or it finishes with a woolly surface.

Balsa is a weak timber but has good buoyancy and insulating values. The lightness of Balsa has led to its being extensively used for floats etc., but since it can absorb a lot of water it is given special wax treatments before being used for such purposes. The limited uses for the timber include thermal and acoustic panelling, aircraft work and toymaking (more especially aircraft modelling). Widely used for insulation in cold stores and refrigerated ships. It is also used for special packaging.

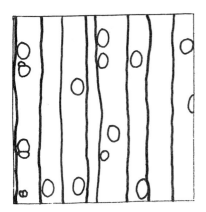

No particular structural features are evident. Identification is normally safe on the basis of the low weight of the wood.

Basswood (American Lime)
Tilia americana

This timber is closely related to the European Lime, which it resembles in appearance but it is the product of a North American tree that may grow to over 30 metres. This timber should not be confused with that of American Whitewood (*Liriodendron tulipifera*) which may also be sold under the name of Basswood. The timber weighs about 420 kg/m³ when dry and is straight grained, with a fine and even texture; it ranges from a creamy white to lightish brown in colour but there is no distinctive figuring.

Movement   –   small
Durability   –   non-durable
Treatability   –   permeable
Workability   –   very easy to work; finishes well

After drying has been completed the wood is remarkably stable and well-suited for such purposes as pattern-making, where freedom from movement is of importance. Its greatest importance is on the American market but it is imported into the United Kingdom. In addition to pattern-making, some typical uses for Basswood include the making of musical instruments, mouldings, picture frames, match splints, boxes, drawing boards, plywood, food containers, turnery and wood wool.

Beech
(Fagus sylvatica)

This is one of the best known and most useful commercial timbers in the world and a steady permanent demand exists. It grows throughout Europe across to Asia. From more northern areas it has an average weight of 720 kg/m^3 when dried; from central Europe it averages 670 kg/m^3. The timber is typically whitish to light brown with an indistinct sapwood. Trees that have grown in the coldest areas may produce logs with 'red heart'; such logs have a pattern not unlike a flower on their ends – this produces reddish-brown streaks in the length of the timber. It is common practice to steam beech in some parts of Europe and this causes a permanent pink or light red colouration.

The wood is straight grained and has a fine and even texture.

Movement	–	large
Durability	–	perishable
Treatability	–	permeable
Workability	–	generally easy but depends on hardness; some burning can occur in machining, producing a blackened surface; finishes well; pre-boring required

Beech is unusual in that it can be steam bent to quite small radii; this makes it particularly useful in the furniture industry. It can be finished easily to match up with almost any other timber. The timber is actually stronger than Oak and because of its straight grain and ease of working can be used for almost anything – it is popular for all types of furniture, joinery, tool handles, brushes, sports goods and flooring. It is also used extensively for plywood.

(continued overleaf)

71

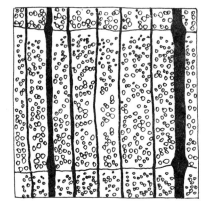

Rays are unusual in being of two sizes. The larger ones show clearly on a flat-sawn surface as darker brown canoe-shaped flecks. Scalariform perforations are present.

Beefwood

Casuarina equisetifolia

The tree providing this timber is one of rapid growth habit that is native to various parts of Australia, South East Asia and adjacent regions. It has an average dried weight of about 960 kg/m^3 with a grain either straight or shallowly interlocked and a texture which is moderately coarse but even. The heartwood may be red or run through various shades of red-brown in colour and, according to the degree of interlock grain present, quarter-sawn stock may show a stripe figuring. The timber is rather difficult to work by hand and is classed as refractory under drying treatments as it is very apt to split, both in air and kiln processes. It is not a naturally durable timber when exposed to the air or when it is in contact with the soil. Although enjoying a certain amount of local importance the timber is never likely to become an important factor on the general world markets. Typical uses for the wood include posts, rafters, mining and structural timbers.

Benteak (Nana)
Lagerstroemia lanceolata

This is a not very well-known Indian species closely related to Pyinma. It is light red or reddish-brown, darkening with age to a uniform medium brown. The surface is lustrous and the grain straight.

The wood is similar to Teak in most respects and has found limited use for furniture, joinery, boat-building and vehicles.

Birch

(geographical names according to origin)

Betula pubescens & *Betula pendula*

Birch is one of the best known of the world's timbers. It is timber from trees of rapid growth but only medium height; they are of common occurrence in the north temperate zone and are often regarded as weeds. The wood is white to a very light brown in colour, with straight grain and a fine and even texture; it weighs about 660 kg/m^3 when dried.

Movement	–	not known
Durability	–	perishable
Treatability	–	permeable
Workability	–	easy to work but can sometimes finish with a woolly surface unless well sanded

The timber polishes satisfactorily and its lack of colour and figuring make it a useful wood for staining to match another wood; it also turns very well and bends satisfactorily. Among the uses to which Birch is put are plywood manufacture, boat and canoe building, fish casks, hoops, shoe heels, kitchen furniture, dowels and chair and cabinet making. In the United Kingdom the Birch tree grows only to an average height of some 10 metres, hence large timbers are not readily available but in Europe solid wood is larger. Whether solid or plywood small brown flecks caused by insect damage to the tree are often seen on vertical surfaces. Unless these pith flecks affect appearance they are not important. Severe attack also causes grain disturbance which is decorative when peeled and is sold as Masur Birch.

(continued overleaf)

Birch *continued*

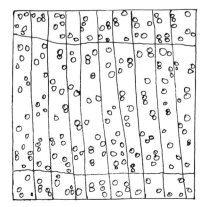

Growth ring boundaries are just visible. Scalariform perforations are present but hard to see because the vessels are so small.

Birch, Paper

(White Birch, American Birch)

Betula papyrifera

As with all the true birches, this timber is the product of a tree of rapid growth habit. It is common in various parts of North America though there is less of it than of Yellow Birch and supplies of the timber are sometimes seen on the United Kingdom markets. In colour it is a uniform creamy-white with no distinctive figuring and is straight grained, with a fine and even texture. Average samples weigh about 620 kg/m^3 when dried (lighter than the two preceding timbers).

Paper Birch is rather easier to work, whether by hand or machine, than the majority of the commercial birches. It is a wood especially suited for turnery work but its main use is for plywood manufacture.

Scalariform perforations are present.

Birch, Yellow
Betula alleghaniensis & *Betula lenta*

This timber grows in Eastern Canada and USA and may reach the market under such names as Quebec Birch, Canadian Yellow Birch or Canadian Silkywood (if strongly figured). It is normally a straight-grained, fine and even-textured wood. It is rather heavier than the English species, the weight averaging 690 kg/m³ when dried and the wood ranges from light to dark reddish-brown in colour.

Movement   –   large
Durability   –   perishable
Treatability   –   moderately resistant
Workability   –   easy to work and finishes well when straight-grained

Much Yellow Birch is made into plywood. It is also used for furniture framing, turned items and for heavy duty flooring in schools, gymnasia and dance halls.

Scalariform perforations are present.

Black Bean
Castanospermum australe

This is extremely hard wood that becomes dry and brittle with age, being refractory under drying treatments and very liable to warping and collapse. It was once well-known on the world's timber markets but as the tree producing it has only a very limited distribution in certain parts of Australia, the available supplies are falling off yearly. The timber weighs about 770 kg/m³ when dried and has straight or shallowly interlocked grain; the texture is rather coarse. The sapwood and the heartwood are sharply distinct; the heartwood is chocolate brown to almost black, with a prominent figuring of greyish brown streaks due to parenchyma tissue around the vessels.

Movement	–	medium
Durability	–	durable
Treatability	–	very resistant
Workability	–	reasonably easy to work and finishes well; can be difficult to glue sometimes

This handsome timber is used in the solid and as a veneer for joinery, furniture and interior fittings. Must be dried very carefully or new stock may have various forms of drying degrade.

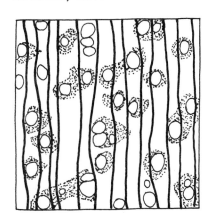

Aliform parenchyma is clearly visible. White deposits are common in many vessels. Ripple marks are present.

Blackbutt
Eucalyptus pilularis

New South Wales and the southern coastal regions of Queensland are where this timber grows. It has an average weight of about 880 kg/m³ when dried. In Australia several closely related and similar species are mixed with it so the weight can be a little variable. The grain is straight or slightly interlocked or wavy and the texture moderately fine. The wood is pale to medium brown, often with some pinkish markings. Small gum veins are quite common.

Movement   – no information
Durability   – very durable
Treatability – very resistant
Workability – fairly hard to work, especially if the grain is not straight; needs pre-boring; finishes well

This heavy timber is used for heavy construction work, fencing posts and rails and sometimes flooring.

Blackwood, Australian

(Black Wattle)

Acacia melanoxylon

The tree providing this timber grows best in Tasmania, though it is also to be found in South Australia, Victoria and on the table-lands of New South Wales. It has an average dried weight of about 660 kg/m^3. The grain is variable, being either straight, interlocked or wavy but the texture is invariably medium and even. In colour the wood may range from golden-brown to a very dark brown, with a figuring of darker streaks and it has a lustrous surface.

Movement   –   believed to be small
Durability   –   durable
Treatability   –   very resistant, including the sapwood
Workability   –   fairly easy to work with straight grain; finishes well

Typical uses for the timber include furniture-making, interior fittings, high-class joinery, gun stocks and cooperage. The timber sometimes seen on the world markets and described as African Blackwood comes from a different species (*Dalbergia melanoxylon*) and is related to the rosewoods.

Australian Blackwood works reasonably well and responds satisfactorily to finishing agents. Selected stock is considered to be among the most decorative of Australian hardwoods and makes an excellent sliced veneer.

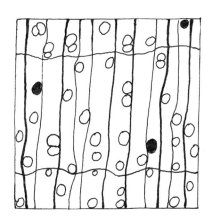

Growth rings are visible as a change in density of the background fibres.

Bloodwood, Red
Eucalyptus corymbosa

The wood is the product of a common Australian tree but the timber is not well-known on the general market. It is a naturally durable timber when exposed to the weather and is especially resistant to fire hazard but the usefulness of the wood is affected by the numerous gum veins to be found in it. Red Bloodwood weighs about 1025 kg/m^3 when dried and has an interlocked grain and a medium coarse texture. It is a uniform red in colouring with no distinctive figuring on flat-sawn surfaces, though quarter-sawn faces may have a slight stripe figure.

Typical uses for the timber include railway sleepers, fencing posts, flooring, telegraph and telephone poles. The wood is unlikely ever to come into common use other than as a local structural or sleeper timber.

Bombway, White (Badam)
Terminalia procera

The timber is a native of the Andaman Islands and is obtainable in very good dimensions; the alternative spelling Bombwe or the Indian name of Badam may occasionally be used. It is a member of a widely distributed botanical genus which includes many better-known timbers such as Indian Laurel, Afara and Idigbo.

The sapwood is not well defined from the heartwood; it is greyish in colour though sometimes blotched with yellow. When freshly sawn the heartwood is a lightish brown shade but it eventually darkens to a greyish-brown; inconspicuous darker markings may be present in the heartwood. Planed surfaces of the wood are mildly silky, the texture is coarse and rather uneven and the grain is usually straight. The average weight for dried stock is about 640 kg/m^3

Movement	–	small
Durability	–	non-durable
Treatability	–	moderately resistant
Workability	–	fairly easy to work; finishes well

This coarse timber needs filling before polishing but is suitable for joinery, panelling, furniture, interior fittings and similar purposes. It can also be used for plywood.

Boxwood   (Abassian, Iranian, Turkish etc. Boxwood)
Buxus sempervirens

This timber is frequently offered for sale under the name of Turkish Boxwood. Although many timbers are sold under the descriptions of Boxwood this is one of the few that really belong to the true Boxwood family. It is the product of a tree common in various parts of Europe and Asia Minor and large quantities of the wood reach the world markets, though it is only available in small sizes. In many places the tree is little more than a bush. The timber is light yellow in colour with a very fine and even texture. The grain is often straight but may be quite irregular, especially in wood from the smallest trees. The wood weighs about 910 kg/m³ when dried.

Movement   –   no information
Durability   –   probably durable
Treatability   –   no information
Workability   –   rather difficult and splits easily; finishes well; needs
                   pre-boring

This timber needs care in drying to avoid surface checks. Its very fine texture and hardness make it a special-purpose timber. It is invaluable for wood engraving because it will accept very fine detail of carving. Other uses include mathematical rules and scales, inlaying, parts of musical instruments, turnery, chessmen, rollers and silk shuttles.

The structure is so fine that even with a hand lens the vessels can barely be seen and growth rings are only faint. Scalariform perforations are present.

Boxwood, Knysna   (Kamassi or Cape Boxwood)
Gonioma kamassi

This species is not a true boxwood but is a very good alternative. It comes from South Africa. It needs to be dried slowly to avoid degrade. It is really in all respects very similar to *Buxus sempervirens* and can be safely used for all the same purposes. It may be available in slightly larger sizes than Boxwood.

The fine dust produced on machining is stated to cause ill-effects such as headaches and giddiness in some people.

Boxwood, Maracaibo   (West Indian Boxwood, Zapatero)
Gossypiospermum praecox

This tree grows particularly in the Maracaibo district of Venezuela. In colour it is sometimes paler than true Boxwood and may be almost white. The name Zapatero is also applied to Purpleheart.

There is a marked risk of fungal staining if the billets of timber are not dried quickly or kept in airy conditions.

The wood is in general very similar in properties and uses to Boxwood except that it is more lustrous and splits more readily.

Bulletwood

(Beefwood)

Mimusops species

This timber, from India, Burma and the Andamans, should not be confused with the wood of *Casuarina equisetifolia*, a very unusual Australian Beefwood. The heartwood is a rich reddish-brown in colour, sometimes with a darker figuring but having a non-lustrous surface. The grain is shallowly interlocked or irregular, the texture fine and even and the average weight for dry stock is about 1040 kg/m³. The timber has a considerable reputation all over the world both for its strength and durability. Bulletwood is not an easy timber to work by hand but unless really deeply interlocked grain is present machining causes little difficulty and the wood finishes well. It takes nails and screws satisfactorily, glues well and is capable of taking a high polish.

Structural work, boat building, furniture and cabinet making, tool handles, wheel spokes and railway sleepers are amongst the uses to which the timber may be put.

Butternut (White Walnut)

Juglans cinerea

This wood frequently reaches the market under the name of White Walnut. It is a North American timber that is a member of the true Walnut genus but is generally regarded as being inferior, as a timber, to American Black Walnut. Like all the walnuts it has an economic importance for its fruits as well as its timbers. It is not an easy wood to season and a high degree of care is necessary to avoid degrade. Butternut is not difficult to work under most hand or machine operations. The timber is brownish in colour but has a lighter tint than the other commercial walnuts, being at times almost greyish. It is also lighter in weight and therefore weaker. The grain is normally straight but the texture is somewhat coarse.

The wood can be used for the same purposes as other walnuts but it is generally regarded as being inferior and is more likely to be used for boxes, crates and small wooden items.

Butternut, Rose
Blepharocarya involucrigera

This species exists in commercial quantities in northern Queensland but the wood is not well-known on the timber markets. It has an average weight of about 650 kg/m^3 when dried, with straight or inter-locked grain and a medium coarse but even texture. The colour is light pinkish-brown or darker but there is no distinctive figuring on plain-sawn faces. It is a strong and tough (but reasonably soft) wood and is said to be expensive to convert as large baulks normally have a black and useless heart. Rose Butternut is not a naturally durable timber and takes a considerable time to dry but once the process is completed it is very stable. The timber works well under hand and machine operations, with little dulling of tool edges and will take most of the normal types of finishing treatment.

It may be used for ordinary domestic furniture, cabinet making, certain types of cooperage and similar purposes.

Camphorwood, East African

Ocotea usambarensis

A large tree growing in Kenya and Tanzania and producing a timber with a distinct camphor scent when fresh. The odour disappears with drying and a little ageing. It has an average weight of 610 kg/m³ when dried, interlocked grain and a moderately fine and even texture. The heartwood colour is yellow with green or brown tinges when fresh but exposure to light turns it dark brown. Slow drying is needed to avoid degrade.

Movement   –   no information
Durability   –   very durable
Treatability   –   no information
Workability   –   easy to work except for planing of interlocked faces

This species is used for furniture, light construction, internal and external joinery and flooring.

Apart from its weight, the timber is a little like its close relative Greenheart. In structure it has slightly more vessels and rather less obvious parenchyma.

Camwood
Baphia nitida

The true Camwood is *Baphia nitida* from West Africa but the name is also applied to *Pterocarpus soyauxii* which is preferably known as African Padauk. To complicate the matter, both timbers are used to produce natural dye by boiling chips in water. They are often used interchangeably where they occur because both are handsome deep reddish-brown woods, often with a slight purplish striping.

Baphia weighs about 670 kg/m³ when dried with a medium texture and *Pterocarpus* about 820 kg/m³ with a coarse texture.

Both timbers are easy to work, finish well and are very durable.

Camwoods are suitable for turnery, cutlery handles, joinery and flooring.

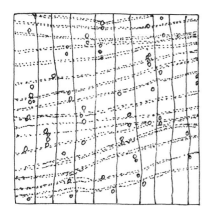

The illustration shows the structure of *Baphia nitida*. *Pterocarpus soyauxii* has much larger vessels and irregular bands of parenchyma.

Canarium, African
Canarium schweinfurthii

(Elemi, Abel, Mwafu)

This is another example of a scientific name being usefully used as a trade name. This tree grows in several West African countries and right across to Uganda. The heartwood of the species is of a pale brown or pinkish-brown colour that sometimes has the appearance of a light mahogany; the sapwood is wide and normally has a pinkish tinge. Planed surfaces of the timber are lustrous, the texture is rather coarse and the grain interlocked. Canarium has a mild, pleasant but not very distinctive smell when freshly cut and weighs about 530 kg/m^3 when dried.

Movement   –   medium
Durability   –   non-durable
Treatability   –   very resistant
Workability   –   not too difficult as long as all the cutting edges are kept sharp; silica is present and this blunts all tools rather quickly; can finish well with care.

Canarium is very much a general utility wood used for interior joinery and carpentry. It produces useful veneer for plywood cores but can be very decorative when the figure is good enough.

Structurally it is a featureless timber; white deposits and tyloses may be present. Large logs may have brittleheart.

Carrobean
Sloanea species

This is a commercially important timber in Australia and is provided by various species of *Sloanea*. Different varieties such as Grey or Blush are recognized but apart from differences in colour tone there is little to distinguish the species. Average weight is about 600 kg/m^3 when dried, the grain is shallowly interlocked and the texture fine and even. The colour is variable but normally a light reddish-brown; quarter-sawn stock shows a silver figure due to the large rays. The wood works well but its durability is not good and it is not especially good in its drying qualities.

It is unlikely to be exported, because many other species offer similar or better properties and are already well-known.

Cedar, Central American (various – see below)
Cedrela odorata

The name Cedar is applied to many commercial species and this list includes both hardwoods and softwoods. The common factor behind this loose use of a name is the presence of oils with the same odour as that of the true Cedar (a softwood).

Central American Cedar is a hardwood of the Mahogany family. It is known by many common names – Cigar Box Cedar is perhaps the best known but there is a long list of geographical names related to the country of origin – it grows almost throughout the Central and South American regions. Cedro and Spanish Cedar are further names that may be used. There is also a South American Cedar (*Cedrela fissilis* principally) to which the same description applies.

The timber varies considerably according to its origin but is generally like a soft mahogany. The heartwood colour ranges from pink through shades of red to quite a dark reddish-brown in different trees. The oil is always present and may sometimes exude onto surfaces. The smell is quite strong, although pleasant, at first, but will virtually disappear with age. If the wood is cut into the smell will reappear. The grain is mostly straight but may be interlocked and the texture is moderately coarse. The weight averages 480 kg/m^3 but varies between about 370 to 750 kg/m^3 when dried.

Movement   – small
Durability   – durable
Treatability – very resistant
Workability – easy to work but needs sharp tools to avoid woolliness; finishes well

The wood has a very mild figuring and is used for good quality joinery, cabinet work, panelling, some boat building and plywood. Cigar boxes or a veneer leaf in an aluminium tube are traditional uses.

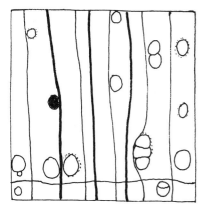

The end-grain usually shows a distinct semi-ring porous structure which produces the obvious growth ring pattern on plain-sawn surfaces or rotary peeled veneers.

Ceiba

(Fuma, Fromager)

Ceiba pentandra

A closely related timber called Bombax is so similar that it is frequently marketed as a mixture. An allied species (*Ceiba occidentalis*) grows in Central America and produces Honduras Cottonwood. *Ceiba pentandra* grows in West and Central Africa. It is frequently the central village meeting tree on account of its great size and is known as the Silk Cotton Tree – its seeds produce commercial kapok.

The wood has an average weight of about 320 kg/m³ when dried and is not very strong; it is chiefly used for secondary purposes such as core stock but locally it may be employed as a softwood substitute for certain types of work. The wood is usually almost colourless or greyish, some samples having slightly darker yellow streaks. Grain is inclined to be rather wild, while the texture is coarse and the planed surfaces have no sheen. If the timber is not converted and dried as soon as possible after felling serious staining may develop.

Movement   –   no information
Durability   –   perishable
Treatability   –   permeable
Workability   –   easy to work but does not finish easily

Apart from core plywoods it can be used for sound insulation, linings, packaging and simple lightweight joinery.

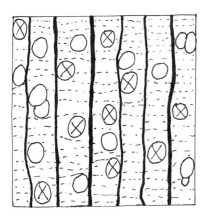

Tyloses are plentiful and obvious. The very fine lines of parenchyma are not so obvious. Ripple marks are present but also difficult to see.

Celtis, African

(Esa, Ohia)

Celtis species

This timber grows in East, Central and West African countries and is composed of three principal species: *Celtis adolfi-friderici, Celtis mildbraedii* and *Celtis zenkeri*.

The sapwood and heartwood can barely be distinguished because the timber is white to pale yellow in colour. Quarter-sawn stock has a ribbon stripe figure due to the interlocked grain, the texture is fine and even and the planed surfaces have a good sheen. Celtis has a mild growth ring figure when flat-sawn. Its average weight when dried is about 780 kg/m³. The timber is difficult to dry without degrade.

Movement   –   medium
Durability   –   perishable
Treatability   –   moderately resistant
Workability   –   moderately easy to work but can be difficult to finish when grain is irregular; needs pre-boring

This wood is very strong and compares favourably with Ash. It is a general utility timber suitable for interior joinery. A hard-wearing timber for flooring.

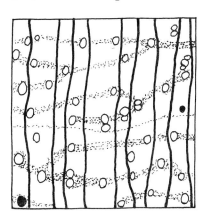

Some vessel deposits are present and the parenchyma is very easily seen although the wood is pale.

Cherry, American (Black Cherry)
Prunus serotina

This is the North American equivalent to the common European Cherry, though it is of greater economic importance as a timber than the European wood. The timber has low shrinkage factors and dries easily. Its straight grain, fine even texture and warm red-brown colour make it very similar to European Cherry but it is slightly lighter in weight at 580 kg/m^3 when dried.

It is also similar in all other respects and is used for furniture, cabinet-making, pattern making, musical instruments and good quality joinery.

Cherry, European

Prunus avium

(Gean, Wild Cherry)

Economically the Cherry is not regarded as a major timber tree although it is well known for certain minor uses and it is not uncommon on the market. The timber has a rather high oil content and will not always polish satisfactorily, though it responds well to most other kinds of finishing treatment. The dried wood weighs about 600 kg/m³.

The grain is straight, the texture moderately fine and even and the heartwood pinkish to light brown in colour. The timber has no distinctive taste or smell. Closely related species include Austrian Cherry (*Prunus mahaleb*) and American Cherry; the tree is related botanically to both the Apple and Pear.

Movement   –   medium
Durability   –   moderately durable
Treatability –   no information
Workability –   works well, depending on irregularity of grain

Exposure to light seems to improve the colour of Cherry and it looks very good in furniture, panelling and decorative joinery. It is used for a variety of small turned items also.

Cherry wood is one of the few semi-ring porous timbers. The growth rings are marked by a zone of numerous earlywood vessels. These vessels are not larger than those in the latewood – only more numerous.

Chestnut, American

(Sweet Chestnut)

Castanea dentata

This timber was once of considerable importance and was imported in some quantity as an Oak substitute. Unfortunately a Sweet Chestnut borer (similar to Dutch Elm Disease) killed most of the trees and it is now in very limited supply.

The average weight is about 480 kg/m^3 when dried and the pale brown heartwood is distinct. The grain is usually straight, the texture coarse and the wood contains tannin.

The wood is very similar to Oak in appearance; it is ring-porous so it shows distinct growth ring figure. There are, however, no large rays present so it does not show silver figure.

American Chestnut is used for furniture, fence posts and railway sleepers. The tannin content causes severe staining when in contact with moisture and iron.

Chestnut, Sweet   (Spanish or European Chestnut)
Castanea sativa

This timber closely resembles Oak in its general appearance but quarter-sawn material does not show the prominent ray figuring, or so-called silver figure (or silver grain), that is so characteristic of the true oaks. The tree producing the timber is common to all parts of the north temperate zone. The heartwood of the timber ranges from a light yellowish-brown to a dark brown in colour; the grain is normally straight and the texture coarse and rather uneven. Average air dried weight is about 540 kg/m³. Occasional logs show severe spiral grain.

Movement   –   small
Durability   –   durable
Treatability   –   very resistant
Workability   –   easy to work and finishes well

Care has to be taken to see that this wood is thoroughly dry before use. This is particularly so because it is slow to part with its moisture and if damp will corrode metals and cause staining when in contact with iron. It is an easy timber to split and there is some demand for conversion into split pale fencing. Other typical uses for the timber include the making of garden and church furniture, gates, panelling, telegraph and telephone poles, and domestic furniture (when it is often stained to represent Oak).

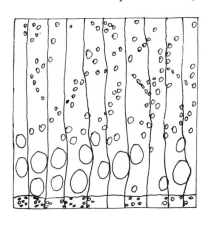

A ring-porous timber with a marked tendency to oblique lines of latewood vessels.

Chickrassy

(Chittagong Wood, Yimna)

Chukrasia tabularis

This Indian and Burmese species produces a golden-mahogany to reddish-brown timber with a pronounced growth ring figure. The grain is irregularly interlocked or wavy and this can lead to further beautiful figuring and lustre. The wood weighs about 620 kg/m^3 and is easy to dry.

Movement   –   small
Durability   –   non-durable
Treatability –   very resistant
Workability –   easy to work and finishes well

It is useful for furniture, interior joinery and carving. There is a slight tendency for fine surface checks to develop during drying; their visibility will be accentuated by polishing.

Coachwood
Ceratopetalum apetalum

This timber has an average dried weight of about 630 kg/m^3, with a light brown to pinkish-brown heartwood (uniform in any one specimen); the sapwood is very indistinct and only a little paler. It has straight grain and a fine and even texture. Coachwood is an Australian timber coming from a tree to be found in various districts of South Australia and Queensland but, although it is available on the world markets from time to time, it is not likely to challenge the supremacy of better known timbers of the same class. It is not a very difficult wood to work, turns well and could be summarized as an extremely useful, slightly decorative cabinet timber. It is not a durable wood. More use might possibly be made of the timber in the United Kingdom for the manufacture of veneers. In Australia the wood has sometimes been described as Rose Mahogany or Scented Satinwood, though it has no valid claim to the title of either mahogany or satinwood. Suitable for furniture, mouldings and flooring. Scalariform perforations are present.

Cocobolo
Dalbergia retusa and allied species

Cocobolo is closely related to the rosewoods and is of Central American origin. Only small quantities of the timber reach the world markets, however, and that is almost entirely taken by the cutlery trade for the making of knife handles and similar items. The wood is naturally resistant to decay but needs careful attention during drying processes if the rate of degrade is to be kept within acceptable bounds. It works reasonably well in all hand or machine operations and is a good timber for turnery purposes. The heartwood is a very variable yellow to dark reddish-brown in colour, with a figuring of darker irregular markings; it has a faintly fragrant smell. The grain is usually straight but may be wavy, while the texture is fine and even. Average dried weight is between 990 and 1200 kg/m³. The wood has a natural cold feel like marble.

Cocuswood

Brya ebenus

This timber has the misleading alternative names of Jamaica Ebony or West Indian Ebony. It is a Tropical American timber coming from a tree of low growth habit and is available only in the form of small logs. The heartwood is very dark brown to blackish in colour and is the only part of the tree that is of any commercial use. Cocuswood causes little trouble in drying, does not readily warp or check and is classified as durable when exposed to the weather or used in contact with the soil. It is somewhat brittle and should be pre-bored before nailing, screwing, etc. but the wood turns and finishes satisfactorily. Typical uses include cabinet work, inlaying, brush backs, turnery, parquet and musical instruments such as flutes and clarinets. It is very heavy (average 1200 kg/m^3 when dried) and has a fine and even texture, with the grain either straight or wavy. Although it is not a well known timber in the strict sense of the word, supplies of Cocuswood are normally obtainable on the United Kingdom markets.

Cordia, African

Cordia species

This timber is usually marketed as a mixture of *Cordia abyssinica*, *Cordia millenii* and *Cordia platythyrsa* which grow right across Tropical Africa.

In colour the wood can be a pale golden brown or medium brown or even pinkish. Its weight is also variable but averages 430 kg/m³ when dried. The texture is rather coarse and the grain is straight to slightly interlocked, or irregular. Careful drying is needed to remove moisture evenly.

Movement	–	small
Durability	–	moderately durable
Treatability	–	resistant
Workability	–	easy to work provided tools are sharp; finishes well

Care is needed to avoid the softer pieces for vulnerable surfaces – otherwise Cordia is quite suitable for furniture and internal joinery. A traditional drum wood in Africa owing to its resonant qualities. Brittleheart sometimes present.

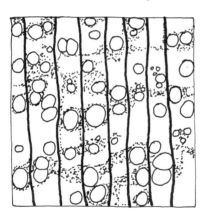

The parenchyma is very variable. It may be almost invisible, vasicentric, aliform or almost banded.

Cottonwood, Eastern
Populus deltoides

Cottonwood is a North American species of poplar that flourishes in the upper parts of the Mississippi and Missouri valleys. Normally the tree is the largest of the poplars and consequently very wide boards are readily obtainable. Its properties are much the same as those of Poplar, that is to say it works easily in all hand and machine operations, carves and turns well and takes stain, paint and other finishing agents satisfactorily. It may cause a certain degree of difficulty in drying, showing a tendency to warp or twist. It is rated as moderately durable and is rather difficult to impregnate with wood preservatives. Cottonwood is lightish in colour, with no characteristic taste or smell, has a fine and even texture and is normally straight grained. It is not particularly heavy, averaging 430 kg/m³ when dried.

The Black or Balsam Cottonwood (*Populus trichocarpa*) is a similar species, native to the Western states of the USA and Canada. Canadian Poplar (*Populus balsamifera* and *Populus grandidentata*) is also typical of the group; the timbers of all of these species are almost identical in their working and other characteristics, though the Eastern variety is possibly the most durable of all the cottonwoods.

Courbaril (West Indian Locust)
Hymenaea courbaril

This timber is best known on the American markets (where it is usually sold as West Indian Locust), being the product of a tall Tropical American tree. In addition to its timber the tree also yields South American copal gum, which is an important item of commerce. Courbaril is not an easy timber to work by hand but is much less difficult by machine and finishes smoothly under a machine planer. It is only moderately durable. It is a hard timber with an average dried weight of 910 kg/m^3, variable as regards grain, with a medium coarse and uneven texture. The sapwood is broad, pinkish to greyish in colour, while the heartwood is orange to dark brown; dark streaks may be present. The wood has a pronounced silver figure on quarter-sawn surfaces. Typical uses for the timber include furniture, cabinet-making, ship-building, carpentry and similar purposes. Courbaril is normally available in timbers of good dimensions; it is a very tough wood so can be used as a substitute for Ash for sports goods and tool handles.

Crabwood

(Andiroba, Krappa)

Carapa guianensis

Central and South America supply the bulk of the world's supply of this wood. In the timber trade Crabwood is qualified as either Red or White according to the predominating hue, but there is no botanical or other essential difference between them. The wood may also reach the market under the titles of British Guiana Mahogany, Brazilian Mahogany, or Demerara Mahogany. It is closely related to the mahoganies but it is best not to use these confusing names. The colour varies from pale pink to red-brown when freshly sawn but darkens to an even dull reddish-brown. The grain is mostly straight but may be interlocked; the texture is rather coarse and the weight about 610 kg/m³ when dried – this is a little variable, however.

Movement   –   small
Durability   –   moderately durable
Treatability –   no information
Workability   –   reasonably easy to work, finishes well

Splits may develop on drying without great care. The wood is not as good as true mahoganies but is suitable for furniture and joinery. Used for construction work generally in Guyana.

The African Crabwood, or Crabnut (*Carapa procera*) is similar to the Central American species in all respects, but is of restricted distribution and limited commercial importance.

Danta (Otutu)

Nesogordonia papaverifera

This species grows in the southern parts of West Africa. It has a sharply defined sapwood and heartwood, the former being of a pale brown colour while the heartwood is of a reddish-brown shade. Quarter-sawn stock has a narrow stripe figure, brought about by the interlocked grain, while the texture is fine and even. The planed surfaces of the wood have quite a good sheen and when dried it has an average weight of about 740 kg/m³. Danta sometimes feels slightly greasy to the touch. The wood needs to dry rather slowly, both in air or by a kiln treatment and shows a slight tendency to warp. The more greasy samples of the wood may sometimes cause trouble either in glueing or polishing, though usually the wood behaves well in such operations.

Movement   –   medium
Durability   –   durable
Treatability   –   resistant, including the sapwood
Workability   –   moderately easy; may pick up with interlocked grain; needs pre-boring

Danta is a strong and fairly heavy timber that is suitable for most types of constructional work. It is used for vehicle bodies and for flooring owing to its good abrasion resistance.

Some red deposits may be present in vessels but the structure all tends to be fine and indistinct. Ripple marks are present.

Degame   (Degame Lancewood, Lemonwood)
Calycophyllum candidissimum

This timber may be known in this country as Degame Lancewood and on the American market as Lemonwood. It is imported from Central and South America, in the form of short poles (Degame spars) which still have their bark on. The wood is unusual in that only the sapwood is used commercially.

This sapwood is broad, dull brownish-white or almost colourless, with a straight to irregular grain and a very fine, even texture. The dried weight averages 820 kg/m^3. The working properties are good (though it may show some tendency to split out in mortising or drilling) and it responds excellently to stain and polish treatments. Stock showing markedly irregular grain may prove troublesome in drying, but straight grained stock is quite satisfactory; after drying the wood is stable.

The wood of Degame is regarded as a substitute for true Lancewood and is used for tool handles, carving, fishing-rods, shuttles, turnery and archery bows.

Diospyros species

Ebonies are generally thought of as black timbers. In practice some of the species have only a tiny pencil sized black zone and some of them are white all through. Macassar Ebony (*Diospyros celebica*) and Andaman Marblewood (*Diospyros marmorata*) are both variegated black and brown timbers. Indian Ebony (*Diospyros melanoxylon*) and Ceylon Ebony (*Diospyros ebenum*) are both black species. African Ebony consists of *Diospyros crassiflora* which is black and *Diospyros piscatoria* which is striped black and brown.

All black ebonies are difficult to dry without splitting or checking. The different species vary in weight but all are heavy; the range is from about 880 to 1170 kg/m^3 when dried. The timber is one of the few that are shipped by weight as opposed to volume.

The grain is usually straight and the texture very fine and even. The timber is strong, hard and brittle where black; all species are difficult to work but finish well eventually and polish excellently.

Ebonies are used almost exclusively for decorative purposes. They turn well and find a large range of applications such as cutlery handles, brush backs, piano keys, parts of various musical instruments and chessmen.

Ekki   (Azobé, Kaku)
Lophira alata

A West African species noted for its durability and usefulness in sea defences. The wood is a dark reddish-brown or chocolate-brown with very obvious white deposits in most of the vessels; the sapwood is pale pink. Interlocked grain is usually present and the texture is coarse and uneven. The average weight is about 1050 kg/m³ when dried.

Movement   –   large
Durability   –   very durable
Treatability   –   very resistant
Workability   –   very difficult except with machine tools; needs pre-boring

Ekki is suitable for heavy construction work such as wharves and jetties, bridges, sleepers and heavy duty flooring. All types of marine work except where its limited sizes favour Greenheart instead.

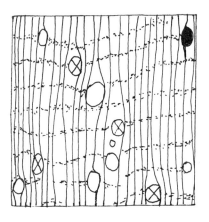

The bands of parenchyma may be narrow or wide.

Elm, English (Red Elm, Nave Elm)
Ulmus procera

The tree producing this timber was common in Northern Europe but in recent years it has been severely hit everywhere by a fungal disease carried by a beetle; the tree is now hard to find. It was normally felled for timber when it was between 20 and 30 metres in height; the tree may normally have a life as long as 140 years. A considerable amount of the trunk is free of branches and in consequence lacks knots. The heartwood of the timber is a dull reddish-brown in colour and average specimens weigh about 550 kg/m³ when dried. The texture of the timber is coarse and uneven, whilst the grain is normally deeply interlocked or irregular. Elm needs considerable care in seasoning as the amount of degrade from warp or shake may be high.

Movement   –   medium
Durability   –   non-durable
Treatability   –   moderately resistant
Workability   –   basically difficult because of the grain and drying distortion; finishes well with care

The distinct growth rings and grain deviations give Elm an attractive appearance which looks well in furniture or veneers. The wood is tough and resilient, far from brittle and is especially noted for its durability under water; it may be used for wagon-making, coffins, agricultural implements, gymnasium equipment, pulley blocks, ship building and turned ware such as bowls. It was formerly used as floor boards.

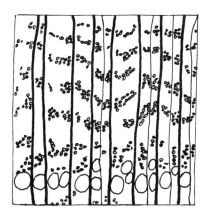

– Elms are all ring-porous and show distinctive banding of small latewood vessels.

Dutch Elm (*Ulmus hollandica* var. *hollandica*) is very similar to English Elm in all respects; it may be preferred because its grain is straighter.

Wych Elm (*Ulmus glabra*) is again very similar except that distinct green streaks often run along the grain and it is somewhat heavier at 690 kg/m^3 when dried. It is preferred for boat work because its grain is straighter than either English or Dutch elm.

Elm, Rock

Ulmus thomasii

(Canadian Rock or Cork Elm)

Rock Elm shows all the toughness and resilience associated with elms and ranks as an important timber on the world markets. Its durability is on a par with that of the other elms and it is described as being a difficult timber to impregnate with wood preservatives. It is heavier than the European species (averaging out when dried between 620 and 780 kg/m^3). The grain is almost invariably straight, whilst the texture is moderately fine and uniform. The sapwood and the heartwood are not sharply differentiated, the wood being lightish-brown in colour with no characteristic taste but a faint, almost imperceptible smell. It is more difficult to work than European Elm but surfaces satisfactorily and successful results may be achieved with steam-bending treatments. Rock Elm has high shrinkage values and is somewhat refractory under drying processes, being subject to warping and checking.

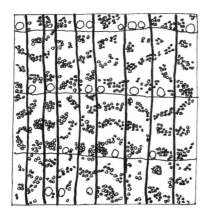

Rock Elm has much less conspicuous ring-porous structure than other elms; large vessels are present but rare so the growth ring figure is not so pronounced.

Elm, White

(American Elm)

Ulmus americana

White Elm may also be given the titles of Soft Elm or Water Elm. It has an average dried weight of about 560 kg/m^3, with either straight or interlocked grain and a coarse texture; the heartwood is lightish-brown in colour. The shrinkage values of this species from North America are lower than those of Rock Elm or European Elm; some specimens may cause considerable difficulty in surfacing owing to the interlocked nature of the grain and a woolly surface may be obtained. Ample supplies of the timber are available to meet normal demands, though the wood is more popular on the American than on the United Kingdom markets. Typical uses for White Elm include flooring, athletic goods, wheelwrights' work, cooperage, agricultural implements, boat building, chairs, ladders and bent-wood work. It is weaker than Rock Elm but otherwise similar. White Elm has a single row of earlywood vessels.

Endospermum (Kauvula)

Endospermum species

Endospermum myrmecophilum is found in Queensland, Australia and *Endospermum medullosum* grows in Papua New Guinea, the Solomon Islands and Fiji.

This timber is of local importance only. It is a pale yellow colour with straight or interlocked grain, a moderately coarse and even texture and weighs about 440 kg/m^3 when dried.

The wood is soft, weak and non-durable and is used for a variety of general purposes to make inexpensive furniture, turnery, flooring and plywood.

Eng

Dipterocarpus tuberculatus

This species comes mostly from Burma but also grows in Thailand. It is of the same genus as Keruing, Gurjun, Yang and Apitong and grows mixed with them in the forests.

A full description which also applies to Eng is to be found under Keruing. The only real difference is that Eng is considerably heavier and therefore stronger. It weighs about 950 kg/m^3 when dried and has more or less the same uses as Keruing.

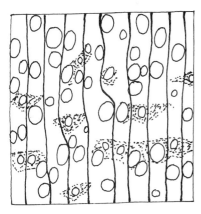

Gum canals are present and the parenchyma development is rather variable.

Esia

(Owewe, Minzu)

Combretodendron macrocarpum

A West African species, especially common in Nigeria. The heartwood is a strong pink colour when fresh but soon darkens to a reddish-brown; darker streaks are sometimes present and a slight ray figure shows on quarter-cut material. The wood has an unpleasant odour on cutting green stock but it disappears totally on drying. The texture is moderately coarse, the grain straight to interlocked and the average weight 800 kg/m³ when dried.

Movement   –   large
Durability   –   durable
Treatability –   very resistant
Workability   –   difficult to work and has a moderate blunting effect

This wood suffers considerable checking and splitting during drying. This makes its utilization difficult other than for construction and railway sleepers.

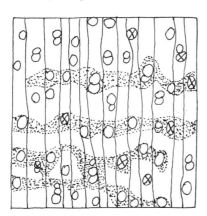

The parenchyma may be aliform or confluent.

Espavel
Anacardium excelsum

A rather undistinguished Tropical American species that finds little use other than for carpentry, joinery and packaging. The wood is a pale pinkish-brown or yellowish-brown, with streaks. Interlocked grain gives a good striped figure but the wood is difficult to finish owing to a natural woolly nature. Added to this, the texture is coarse, requiring filling to polish it. Espavel weighs about 530 kg/m^3 when dry but is weak for its weight and much larger sections are needed to compensate for this. It is unlikely to become more popular than for purely local use.

Freijo
Cordia goeldiana

This timber is sometimes described as Brazilian Teak or Brazilian Walnut. It is a species of the Amazon basin and Tropical America and it is neither a true teak nor a true walnut. Some samples of the wood do resemble Walnut, the heartwood being brown or of a greyish shade of brown with an irregular striped figure. The grain is straight and the texture moderately fine and uneven. Freijo weighs, on average, about 590 kg/m³ when dried. The wood dries well and is stable when dry. Slight borer damage may occur with the timber but otherwise it is durable. Working qualities are good though it is necessary to use really sharp tools to avoid a woolly finish: the timber may be stained and polished after grain filling. Freijo is used for furniture, cabinet work, interior and exterior joinery and fittings and boat-building. It is also useful for cooperage when the quality is good enough.

Gaboon

Aucoumea klaineana

(Okoumé, Gaboon Mahogany)

This timber is not botanically related to the true Mahogany. It is from Equatorial Guinea and parts of French Equatorial Africa. Good-sized logs of the timber are available and it is a favourite species for veneer and plywood manufacture but it may also be used for furniture and interior fittings. The narrow sapwood of Gaboon is lighter than the heartwood which, when freshly sawn, is of a rather distinctive salmon-pink colour, though it will eventually tone to a pale red-brown. Planed surfaces of the wood are silky; the medium texture is rather uneven, while the grain is slightly interlocked. It is a lightweight species, averaging (when fully dry) only 430 kg/m³. This is a non-durable timber both as regards fungal and insect attack and it is resistant to treatment.

Gaboon contains silica and has a marked blunting effect on tools. Since most of this species appears as plywood this fact is not too important. A woolly surface may be present on some of the softer sheets of plywood and this may need scraping to get a good finish. Brittleheart is sometimes present.

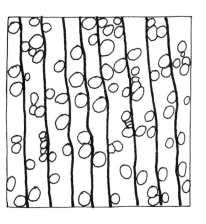

This timber has a very plain structure which is easily distinguished from the true mahoganies and their close relatives. No vertical parenchyma is present.

Gamari (Kumbar)

Gmelina arborea

This Indian and Burmese timber somewhat resembles Teak but is lighter in weight and not so coarse grained. It has an average weight of about 480 kg/m³ dried, a medium coarse, or coarse and uneven texture and an interlocked grain. The sapwood of the timber and the heartwood are not sharply differentiated, the wood being light yellowish-brown in colour with no distinctive figuring but with an oily feel. It is not difficult to work and it responds satisfactorily to most of the normal types of finishing treatment. It is rated as durable. In drying, the wood creates few problems although it shrinks a lot; it is stable after drying. It is less strong than Teak (to which it is related) but is still strong for its weight.

Although native to Southern Asia it is widely grown as a plantation species in West and South Africa. It grows very fast and can be unusually knotty for a hardwood if not well managed.

The wood is suitable for light construction, flooring, furniture, boat-building, joinery and plywood manufacture.

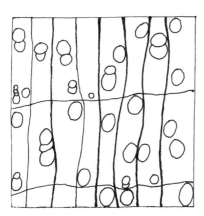

Vessel size will be small in older material and larger in plantation produce.

124

Gedu Nohor (Edinam, Tiama)
Entandrophragma angolense

This timber is closely related to several other commercially important African timbers. It is a member of the true mahogany family. The heartwood is an even reddish-brown and the very wide sapwood is pinkish-grey; occasional logs are much lighter in colour and look like sapwood all through. The texture is moderately coarse and the grain irregularly interlocked; the figure is very uneven and the stripes generally too broad to be attractive so the wood is not as popular as is its close relative Sapele. Distortion on drying can be quite severe. The weight is about 540 kg/m^3 when dried.

Movement	–	small
Durability	–	moderately durable
Treatability	–	very resistant
Workability	–	fairly easy to work except planing; finishes well with filling

This wood is used as an alternative to Mahogany for furniture; it is also suitable for interior and exterior joinery, for vehicle building and some boatwork.

The vessels are fewer and larger than those of Sapele and the parenchyma bands are quite broken and wavy. Ripple marks may be present.

Gonçalo Alves
Astronium fraxinifolium

This wood is known by various common names (Locustwood, Kingwood, Courbaril and Zebrawood) which are also used for other timbers and so can be misleading.

This Brazilian species is a rare but important timber that is extremely popular for the making of high-class furniture and similar purposes. It has an average dry weight of 950 kg/m^3, a grain that is often irregular or interlocking and a medium texture. The light-coloured sapwood is well distinguished from the heartwood, which may range in colour from a lightish-brown to a red tint, with darker stripes. It is naturally resistant to decay, though this is of no great importance as it is chiefly used for internal work. The working and drying qualities are a little difficult. Apart from furniture it is used for small fancy goods and turnery.

Greenheart

Ocotea rodiaei

Greenheart is a timber from Guyana in South America. It is very heavy, average specimens of the wood weighing about 1030 kg/m^3 when dried. The grain is straight, the texture fine and even, and the heartwood is light to dark olive green in colour with a darker streaky figuring; the sapwood and the heartwood are not always sharply differentiated. The colouring is quite variable but this appears to have no connection at all with any of the other properties such as durability. It dries slowly and with some degrade in the way of splits and checks rather than distortion.

Movement   – medium
Durability   – very durable
Treatability   – very resistant
Workability   – moderately difficult to work and blunts cutting edges rather quickly

This timber has particularly good strength properties and is well known for its good resistance to attack by marine borers. For these reasons it is extensively used for all kinds of marine constructions – sea defences, piers, jetties, docks, locks and boat construction. It is also used for heavy construction such as bridges, outdoor decking and engine bearers. Some use is also made of it for fishing rods and other small purposes.

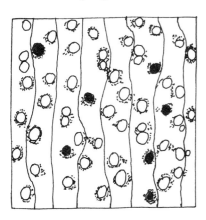

Gum deposits are usually present.

Guarea   (Bossé, Scented Guarea, Black or White Guarea)
Guarea species

The names Nigerian Pearwood or Nigerian Cedar have also been recorded. Imported mostly from Ghana and Nigeria there are two species available: *Guarea cedrata* and *Guarea thompsonii* are normally sold mixed. The timber has a pleasant cedar-like smell when fresh and this may persist for some time. The timber weighs about 600 kg/m³ or a little more, after drying. The colour is pinkish-brown to medium brown at first but becomes more of a light to medium golden brown on exposure to light; basically it is similar to a pale African mahogany but has a finer texture. Sometimes a little gum may be present and can exude under warm conditions; this is not common. The texture is fine.

Movement   – small
Durability   – very durable
Treatability – very resistant
Workability   – fairly good but can dull cutting edges because *Guarea cedrata* contains silica

The timber is related to the mahoganies but is variable in grain direction; sometimes straight it can also be wavy or interlocked. Some people have experienced skin irritation from the dust produced in working – the solid wood gives no trouble.

Guarea is generally of a plain but rich colour suitable for furniture, interior fittings and good quality joinery. It is also used for vehicle building and as a flooring suitable for use with underfloor heating.

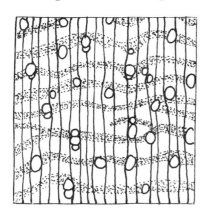

The banded parenchyma tends to be wavy in *Guarea cedrata* and straight in *Guarea thompsonii*. *Guarea cedrata* has more vessels and rays per unit area.

128

Gum, American Red

(Sweet Gum)

Liquidambar styraciflua

This is one of a small group of timbers that may be sold under different names according to whether sapwood or heartwood is supplied. The whitish sapwood (which occasionally has a pinkish tinge) may be marketed as Sap Gum, while the name of Red Gum is given to the heartwood, which varies in colour from brown to reddish-brown, though often having a strong figure of darker streaks. The name of American Red Gum is, however, to be preferred for the timber.

The texture is uniformly fine and surfaces finish with a good lustre. The grain is often irregular and the wood weighs about 560 kg/m³ when dried. This wood needs careful drying and tends to distort if subjected to much atmospheric moisture change in use.

It is used widely in the USA for furniture, joinery, doors, plywood and packaging. The structure as a whole is so fine that little detail is visible, even with a lens. Scalariform perforations are present.

Gum, Spotted   (Lemon Scented Gum, Macula)
Eucalyptus species

This Australian Wood is provided by a mixture of two species – *Eucalyptus maculata* and *Eucalyptus citriodora*. It is also grown on plantations in South Africa.

The wood is highly variable in colour, ranging from light greyish-yellow through greyish-brown to dark brown. The texture is uniformly coarse and the weight about 1000 kg/m³ when dried. Grain is straight or slightly interlocked and the wood feels somewhat greasy. This moderately durable timber is very hard and strong and is a good structural timber for bridges and general work, vehicles and decking.

In Australia the title 'Gum' is applied to other *Eucalyptus* species, the most important being the Blue Gum (*Eucalyptus globulus*) and Saligna Gum (*Eucalyptus saligna*). Both are hard and heavy timbers, well suited for structural work. Many eucalypts are grown in plantations; such timber tends to be lighter in colour and weight. It may well be weaker too.

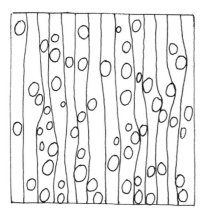

A common structural feature of eucalypts is the presence of more or less oblique lines of vessels as seen on the end-grain (see Jarrah).

Haldu

(Kwao)

Adina cordifolia

Haldu is a timber of Indian origin but the same species may be exported from Thailand under the name of Kwao. The timber is of more than local importance and is considered to be one of the most useful of the timbers native to India.

The sapwood and heartwood of Haldu are not clearly defined, the former being wide and yellowish-white and merging only gradually into the yellowish heartwood. The grain is almost invariably straight, the texture fine and even, while the planed surfaces of the wood are quite silky. The average weight in the dried condition is about 650 kg/m^3.

Haldu works well in all hand and machine operations. It is non-durable but if necessary it can be readily treated with preservatives. The wood seasons easily. The timber from India is usually considered to be of better quality than that from Thailand.

It is used for furniture, turnery, fittings, brushbacks, toy-making and similar purposes and locally it is even used for structural work. Wears well as flooring. Logs are of good form and dimensions.

Hawthorn
Crataegus species

Hawthorn is most frequently seen as a small shrub of one of several *Crataegus* species but sometimes it develops into tree form, though even then it is slender and its height rarely exceeds 3 to 5 metres, though some species may, under very favourable conditions, reach to twice that height. The timber, therefore, is not economically important but it finds minor use for inlaying and the making of tool handles. Hawthorn is rather difficult to work by hand but carves satisfactorily and responds well to the normal kinds of finishing agents. It is naturally durable for external use and dries easily, being stable once drying is complete. The heartwood of the species is whitish in colour, or may have a yellowish tinge but the timber has no characteristic taste or smell. Normally the grain is straight and the texture moderately fine and even, with average timbers weighing about 650 kg/m^3 when dried.

Hazel
Corylus avellana

Economically the Hazel tree is more important for its nuts than for its timber, so the timber is not to be found on the market in any quantity, although it is sometimes to be found in the form of machine-cut or embossed carving or mouldings; it may also be used for tool handles and certain types of cooperage. It is straight-grained with a fine and even texture. The sapwood and the heartwood are not well defined, the wood being whitish to light reddish-brown in colour, with no distinctive taste or smell. It is an easy timber to work in all processes, finishes with a smooth surface and the worked edges remain sharp; it should be pre-bored before such operations as nailing and screwing and care is also needed in mortising and drilling as there is a tendency towards chipping out. Only a limited demand exists for the wood and supplies should be more than adequate. The weight of the timber varies around 720 kg/m^3 when dried. The American species, *Corylus americana* does not grow to as good a height as does the European. Scalariform perforations are present.

Hickory
Carya species

(Pignut, Mockernut, Shellbark or Shagbark Hickory)

Commercial Hickory is exported from Eastern USA and Canada; it is a varying mixture of four species of *Carya* which are indistinguishable from each other. The four are *Carya glabra* (Pignut Hickory), *Carya tomentosa* (Mockernut Hickory), *Carya laciniosa* (Shellbark Hickory) and *Carya ovata* (Shagbark Hickory). Hickory is a species where the sapwood is preferred for its almost white appearance – this is sold as White Hickory; the heartwood is an undistinguished brown or reddish-brown and is sold as Red Hickory. The sapwood closely resembles the timber of Ash and like that timber it is exceptionally tough and strong. The grain is straight, the texture coarse and uneven and the dried weight about 820 kg/m³ (rather variable).

Movement   –   no information
Durability   –   non-durable
Treatability   –   moderately resistant
Workability   –   works with moderate ease except for the heavier pieces; finishes well but blunts cutting tools badly.

The rate of growth of this timber greatly affects its strength; material with rings wider than 1.5 mm is preferred. The great strength fits it for handles of striking tools and for sports goods of most types. Also used for wheels, ladder rungs and vehicle bodies and for bent furniture parts. It is as good as, or better than, Ash and suitable for all the same uses.

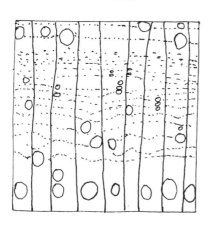

A ring-porous timber which is weaker if the rings are too narrow because there is too little fibrous latewood. Fine lines of parenchyma are distinctive.

Holly
Ilex aquifolium

This wood is used only for items of minor importance such as inlaying, marquetry work, small fancy articles and turnery. It is a hard timber with a dull white or greyish coloured heartwood and an average dried weight of about 800 kg/m^3, an irregular grain and a very fine and even texture. The slow-growing but long-lived trees producing the timber are common enough in the north temperate zone but the wood is available only in fairly small sizes. Holly is not a naturally durable wood when exposed to the weather but is immune to powder-post beetle attack on account of the smallness of its vessel openings. Ample supplies of the wood exist to meet the very limited demand but more use might possibly be made of the timber for such items as wood sculpture and carving for which its very close texture would seem to suit it; it is also useful for fine cabinet work. Working qualities are satisfactory, as it surfaces well and takes a good polish. the wood dries best when cut into small sections ready for utilization. Scalariform perforations are present.

Holly, American
Ilex opaca

This timber, the North American equivalent to European Holly, is a slow growing tree of medium growth habit that reaches its best development in coastal regions. It is not a naturally resistant timber to the attack of wood-rotting fungi but it is immune to powder-post beetle attack. The dry timber is rather lighter than the European species (720 kg/m^3) and is reasonably easy to work, both in hand and machine processes; it finishes well and responds satisfactorily to decorative treatments but it uses are somewhat limited. It is sometimes sold in the form of thin panels for fretworkers' uses and may be marketed with other *Ilex* species under the general description of Holly with no geographical prefix. The heartwood of the species is whitish or light-yellowish in colour, the texture is fine and even and the grain either irregular or wavy. The structure is apparently indistinguishable from the European species. Scalariform perforations are present.

Hornbeam
Carpinus betulus

This timber is by no means a rare one on the world markets, although it is not always available in large sizes. It weighs about 750 kg/m³ when dried with the grain mostly irregular; the texture fine and even; it is whitish or very light in colour. Hornbeam is a perishable wood when exposed to the weather or if in contact with the soil but in view of the uses to which it is put its durability is of minor importance. The working qualities of the wood vary according to the degree of irregular grain present but in general the timber may be said to be difficult to work with hand tools; it finishes very smoothly and polishes very well. Its uses are confined to inlaying marquetry and the making of small novelties, whilst in some types of work the wood may be dyed black and used to simulate Ebony. The American Hornbeam (*Carpinus americana*) is of lower growth habit and has less importance than the European variety.

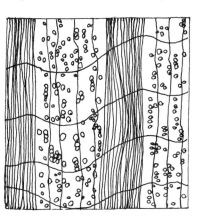

This is one of the very few species to show aggregate rays – a close gathering of ordinary rays that look like occasional very large rays.

Horse Chestnut, European
Aesculus hippocastanum

Horse Chestnut is whitish or yellowish in colour, with the heartwood and sapwood not sharply defined and the timber is very likely to be confused with that of Lime, Willow or Poplar. It is the timber of a common European tree of rapid growth but the wood is not normally available in large sizes and its economic importance is insignificant. It is not related to, nor does it resemble, Sweet Chestnut. The grain of the timber is often crossed or wavy but the texture is fine and even; average weight borders on 510 kg/m³ when dried.

Drying needs to be rapid to retain the whiteness as much as possible. Horse Chestnut is perishable and is mainly used as a cabinet wood. It is an easy timber to work and may be used for such minor items as inlaying and the making of small fancy articles. It is highly suitable for fruit storage trays. Several *Aesculus* species exist on the American continent but are of no importance as timber.

Idigbo

(Emeri, Framiré)

Terminalia ivorensis

A West African species which is closely related to Afara; it is sometimes called Black Afara but this is a name for the tree and should not be used for the timber. The sapwood and heartwood of the species are not clearly differentiated, the wood being a pale yellow-brown in colour and, when plain sawn, having a very prominent growth ring figure. Planed surfaces of the wood are moderately silky; the texture is rather coarse and uneven, while the grain is straight or only shallowly interlocked. The average dried weight may be in the region of 540 kg/m³ but this is very variable.

Movement   –   small
Durability   –   durable
Treatability   –   very resistant and also the sapwood
Workability   –   easy to work except for planing quartered stock; best to pre-bore; finishes well if filled

This tree has a light-weight central core and timber from this part often has brittleheart. The wood contains a dye which may stain other materials when wet; it also contains tannin which causes staining when in contact with iron under moist conditions.

Typical uses for Idigbo include furniture, good quality joinery and domestic flooring.

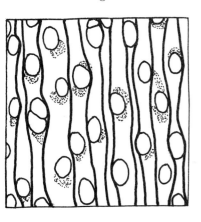

The parenchyma in this species is much less visible than in Afara but otherwise they are very similar in structure.

Imbuia

Phoebe porosa

Alternative spellings of the timber name include Embuya and Embuyia; the title of Brazilian Walnut has also been applied to the species but this is misleading – it is not a true walnut.

The tree is native to Southern Brazil. The sapwood is yellow or yellowish-brown and is not clearly marked off from the heartwood, which is almost colourless when freshly cut though it grows to a light walnut shade with a figuring of lighter and darker streaks. Grain tends to be variable but the texture is fine to medium and even. Average dry weight is about 660 kg/m^3.

Some care is needed in the seasoning of Imbuia to prevent undue warping. Few problems arise in working the wood, though some workers find that the sawdust has irritant properties. The response to finishing agents is satisfactory. Imbuia is used for a wide range of purposes including flooring, furniture and high-quality interior joinery. Useful decorative veneer and plywood can also be manufactured from the species.

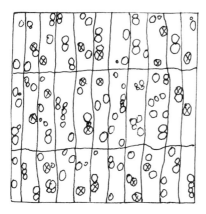

Tyloses are common.
Parenchyma may be difficult to distinguish.

Iroko

(Odum, Mvule)

Chlorophora excelsa

This species is widely distributed through East, Central and West Africa; it has about twenty common names recorded for it, according to country or even district of origin – Kambala, Tule, Intule, Moreira and Bang are among these. Iroko has also been sold as African Teak but this is misleading. *Chlorophora regia* is a West African species so similar that it is not distinguished separately.

When freshly cut Iroko is often quite a strong yellow colour but on exposure it quickly assumes a medium golden brown shade. The parenchyma, which is plentiful in this timber, causes lighter patches to show along the vessel lines. The sapwood is distinct. Overall the timber has quite a close resemblance to Teak but its texture is coarse and even and its grain is either interlocked or irregular. Occasionally large deposits of a hard form of calcium carbonate called 'stone' are present either as lumps or sheets. The dry wood weighs about 640 kg/m^3.

Movement	–	small
Durability	–	very durable
Treatability	–	very resistant
Workability	–	easy to work in general but grain direction or deposits can cause problems; finishes well with filling.

This wood has often been regarded as a substitute for Teak but it is a perfectly good timber in its own right. Its properties in some ways are better than those of Teak. It is used for internal and external joinery, furniture (sometimes as solids combined with Teak veneers), boat and vehicle building, marine uses and flooring (domestic only).

(continued overleaf)

Iroko *continued*

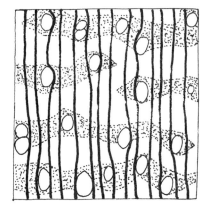

Iroko displays particularly well-developed confluent parenchyma.

Ironbark
Eucalyptus paniculata and allied species

This is a tough, hard and strong timber of the Australian continent that is well known on all markets of the world, being noteworthy for its exceptional durability. It is composed of about 5 species, close botanical relatives of the very well-known Jarrah. Its dry weight is between 1020 and 1180 kg/m^3, the grain is interlocked and the texture fine and even. The wood may range from greyish-brown to reddish-brown in colouring and has a characteristic figure. Ironbark, sometimes called Grey or Red Ironbark, is not easy to work and offers considerable resistance to such operations as nailing and screwing. It causes considerable blunting of cutting edges and the heartwood is classified as impermeable to treatment with wood preservatives. It is a refractory timber in drying processes.

Ironbark is suitable for many outdoor uses, whether heavy construction like bridges, or fencing and railway sleepers.

The wood structure is very like that of Jarrah except that gum deposits are normally absent.

Ironwood
Various genera and species

This a popular vernacular name to describe very hard and tough timbers but to differentiate between them geographical prefixes are often used. Some of the most important varieties are dealt with below.

Canadian Ironwood is the timber of *Ostrya virginiana*, a wood with an average dry weight of about 800 kg/m³, a shallowly interlocked grain and a texture that is medium fine or fine and even. The tree producing the timber is closely related to the Hornbeam and its preferred name is Hop-hornbeam; Canadian Ironwood does not rank as important as Hornbeam on the world markets. It is a very difficult wood to work by hand and is by no means easy to machine but may be surfaced satisfactorily. The wood is classed as naturally durable.

Ceylon Ironwood is a native of Ceylon, India and Malaysia. It is the timber of *Mesua ferrea*, a slow-growing tree of medium height producing a darkish-red coloured wood. The timber is very hard and is naturally durable for exposed positions but is of limited local use only.

East Indian Ironwood is the timber of *Metrosideros vera*, small quantities of which reach the United Kingdom markets. In addition to its use as a timber, East Indian Ironwood has certain medicinal uses. Under drying treatments the timber may prove refractory to handle. The timber is an extremely difficult one to work, whether by hand or machine, has a marked dulling effect on tools but has the merit of being naturally durable. It uses are limited to those purposes where its hardness and toughness are an asset. No special demand is likely to arise for the wood.

Uganda Ironwood is not a commercially important timber but is one that might gain great popularity if it were more readily exploitable. It is a fine-textured, red-brown wood with an interlocked grain, having such a high standard of durability that it is regarded as well-nigh indestructible. For drying it is said to be refractory and serious shakes and splits may develop during the process. Like all timbers of the same name it is difficult to work, whether under hand or machine tools but can be brought to a good surface. It is the timber of *Cynometra alexandri*, although much better known as Muhimbi. It is used for flooring.

Black Ironwood, or South African Ironwood, is the product of *Olea laurifolia* and is therefore closely related to the common Olive. The tree may grow to a height of 20 metres and the timber is rather difficult to dry, usually checking badly. Black Ironwood has a straight grain and a fine and even texture, with the sapwood and the heartwood sharply differentiated – the sapwood being a dull white and the heartwood brownish with a figuring of black streaks. The timber is naturally durable and difficult to work, having a marked dulling effect on tool edges. It is important locally but is never likely to achieve an important place on the general market.

Ivorywood
Siphonodendron australe

The tree producing this Australian timber is not common and the wood is accordingly scarce but when it is available it is prized for such purposes as engraving, turnery, mirror frames, inlaying and fancy articles. It works well in all hand or machine tool processes with little dulling effect on cutting edges. Ivorywood is not naturally durable but is not of the class of woods customarily given a wood preservative treatment. Under drying treatments it is apt to prove rather refractory and is somewhat subject to seasoning stain. Warps and splits are rather common in the converted stock, which is usually found to be rather brittle. Ivorywood is white in colour with a very indistinct growth ring figuring and weighs about 850 kg/m³ when dried. The grain is normally straight and the texture fine and even.

Another species known as Ivorywood is a native of the Argentine, Brazil and other parts of Tropical America and comes from *Balfourodendron riedelianum* but the timber is of local importance only. The sapwood and the heartwood of the species are not sharply delimited, the wood being a uniform whitish or pale yellow in colour, with a non-lustrous surface, no figuring and no characteristic taste or fragrance. Average timbers weigh about 750 kg/m³ when dry, are straight-grained and have a fine and even texture. It is not a naturally durable timber for external use and it works very well, surfaces satisfactorily and may be stained, polished etc. with very good results.

Jarrah
Eucalyptus marginata

Of all the Australian timbers Jarrah is probably the best known and the most popular on the world markets. It may range from pinkish to a dark rich red in colour though the tint is uniform in any one specimen; small dark flecks along the grain are sometimes present. The weight is variable, average dried specimens weighing anything from 690 to 1040 kg/m³; the texture is moderately coarse but even, while the grain is variable.

Movement	–	medium
Durability	–	very durable
Treatability	–	very resistant
Workability	–	rather difficult in all operations

Flooring, heavy construction, cabinet work, panelling and sea defences are among the uses to which the wood may be put.

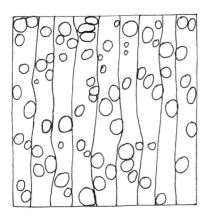

This species is easily confused with Karri – the burning heartwood splinter test is useful to distinguish between them; Karri burns to give white ash and Jarrah burns to give black charcoal.

Jelutong
Dyera costulata

Jelutong is a Malaysian timber of more than local importance and is obtainable in good dimensions; the usefulness of the wood may be marred for certain specific purposes by the latex ducts which are always present. These are slit-like openings on the longitudinal surfaces and travel in a radial direction, often containing dried-up strips of latex which pull out.

The timber is almost white in colour, though the wood will darken slightly in colour after being exposed to the air for a long time; the planed surfaces of the wood are quite lustrous. The grain is almost invariably straight and the texture moderately close and even. Jelutong is light-weight wood, averaging only 460 kg/m³ when dried.

Movement	–	small
Durability	–	non-durable
Treatability	–	probably permeable
Workability	–	easy to work and finishes well

The latex ducts or passages appear in clusters at intervals along the trunk, so any longer piece will invariably show them. The wood is best used in small sizes on account of this. Its stability is particularly good so it is used for pattern-making, drawing boards, carving and as core stock for certain plywoods and doors.

Small latex canals can be seen with a lens in some of the rays as they appear in section on plain-sawn surfaces.

Jequitiba

(Jequitiba Rosa)

Cariniana species

These trees are of tall growth habit and the wood averages about 580 kg/m³ when dried. Straight-grained, the timber has a fine to medium and even texture. In colour the wood is pinkish-brown or light red, sometimes with darker streaks but the figuring is not particularly striking. The timber is easy to work but sharp cutters are necessary to avoid any woolliness of surface.

The two species which make up this timber are not normally separated. They have been sold under the name Brazilian or Columbian Mahogany but they are not true mahoganies at all.

The wood is good for general construction, cabinet-making and boat-building as an alternative to Mahogany.

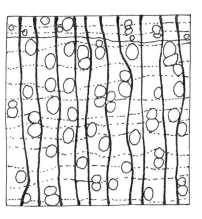

The very fine lines of parenchyma are quite difficult to see clearly.

Kapur

Dryobalanops species

(geographical variations)

Kapur is the timber of about seven species of *Dryobalanops* of Borneo, Sumatra, Sarawak and Malaysia. Various combinations produce Malaysian, Sarawak, Sabah and Indonesian Kapur.

The wood has an average dry weight of between 700 and 800 kg/m³, with the grain either straight or slightly interlocked and the texture fairly coarse but even. The colour of the heartwood is light red-brown or red-brown and the wood has a characteristic smell of camphor when fresh but this soon disappears.

Movement   –   medium
Durability   –   very durable
Treatability   –   very resistant
Workability   –   rather difficult to work but finishes well

The wood stains when in contact with iron under moist conditions. Used for external joinery and light construction; also for furniture and some boat building.

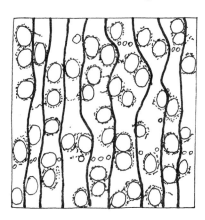

Resin ducts are present in lines but the wood does not bleed. Ripple marks are present.

Karri
Eucalyptus diversicolor

This wood ranks next to Jarrah as the most popular of the Australian timbers exported to the world markets. The two timbers resemble each other but may be distinguished by the 'burning splinter' test in which a sliver of wood about the size of a match stick is burned; a white ash indicates that the timber is Karri and a black charcoal residue that the wood is Jarrah.

This timber has an average weight when dry of about 880 kg/m³. Its grain is either straight or very shallowly interlocked, with the texture moderately coarse but even. The colour ranges from pinkish to dark red but is uniform in any one specimen.

Movement   – large
Durability   – durable
Treatability   – very resistant
Workability   – fairly difficult to work – more blunting to cutters than is Jarrah and fairly difficult to nail

Suitable for much the same uses as Jarrah: heavy construction, cabinet work and bridges. Not so suitable for use in contact with water such as sea defences.

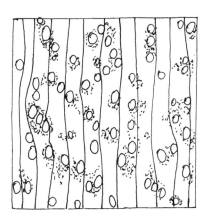

The typical oblique line of vessels common to most *Eucalyptus* species are obvious.

Kempas
Koompassia malaccensis

Kempas is a heavy timber (averaging about 880 kg/m^3 when fully dry) that is well suited for structural work, heavy duty flooring and similar purposes. It finds its greatest importance in Malaysia where it grows. The sapwood and the heartwood of the timber are quite clearly marked off, the sapwood being of a light yellow colour, while the heartwood is a shade of orange-red with yellowish-brown lines when mature, though it is of a pink colour when freshly sawn. The wood is unusual in that some samples show zones of what appears to be sapwood set in the heart, though it is not true sapwood. These zones are hard like a stone and may give rise to fissures during drying. Brittleheart may also be present. It is an acid wood and attacks iron under damp conditions. The grain is interlocked and the texture is rather coarse. Green stock has a very strong smell and this may persist (though much less pungently) in dried timber. Kempas shows conspicuous aliform parenchyma and strongly developed ripple marks.

Keruing (Gurjun, Yang, Apitong)
Dipterocarpus species

There are over seventy species of *Dipterocarpus* growing in South East Asia – many different ones in each country. The species are all broadly quite similar and are marketed as mixtures available under a single name. The name Keruing is the best known in the United Kingdom and actually refers to timber imported from Indonesia, Malaysia, Sabah and Sarawak; Gurjun is timber from Burma and India; Yang is from Thailand and Apitong is from the Philippines. The mixture in any consignment will bring about some variability but it is not normally important; Gurjun and Yang are each composed of a smaller number of species and are therefore the least variable.

The timbers vary in colour from pinkish-brown to dark brown and look rather characterless. Grain is straight or shallowly interlocked and the texture is moderately coarse and even. All of the species contain oleo-resins and many of them will exude it onto surfaces during drying or when exposed to heat or sunshine when in use; gums may also cause problems in machining. The weight is variable too but is generally within the range 720 to 800 kg/m^3 when dried.

Movement	–	medium to large
Durability	–	moderately durable
Treatability	–	moderately resistant at least
Workability	–	fairly difficult owing to grain and gum

All the species tend to be used in place of the more expensive Oak for heavy construction, decking, vehicle building and sleepers. Can also be used for plywood.

(continued overleaf)

153

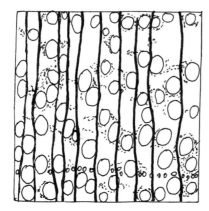

The amount of obvious parenchyma is variable. Gum ducts or canals may be single, in groups or in long tangential lines.

Kokko

Albizia lebbek

(Siris, East Indian Walnut)

This species grows in India, where the name of Siris is preferred, the Andaman Islands and Burma. The name East Indian Walnut is confusing and should not be used – it is not a true walnut. The wood weighs about 650 kg/m³ when dry and has a medium brown colour with attractive markings due to lighter and darker streaks. It also has interlocked grain which adds to the figuring. The texture is coarse but even.

The timber is slow to dry and suffers some degrade; it is also a little difficult to work and needs filling to finish well. It slices easily to yield a highly decorative veneer and this is the form in which it is mostly seen in the United Kingdom. It has been used as a solid for furniture in the past; in India it is used widely for interior building purposes and some boatbuilding.

The wood is almost identical in structure with Albizia from West Africa.

Krabak (various)

Anisoptera species

This timber originates in Thailand as the product of various species of *Anisoptera* but almost identical timbers are available from Burma, Malaysia and the Philippines under the names of Kaunghmu, Mersawa and Palosapis respectively. The different species are mixed commercially and are all similar. The heartwood is yellowish-brown darkening on exposure to a golden brown. The grain is mostly slightly interlocked with a moderately coarse texture. The range of species is variable in weight but averages 640 kg/m^3 dried. The wood dries slowly and, usually, unevenly.

Movement   –   medium
Durability   –   moderately durable
Treatability –   moderately resistant
Workability   –   fairly easy to work but severely blunting to cutting
                  edges

The timber is suitable for general construction, interior joinery and vehicle building; it is also made into a low-grade plywood.

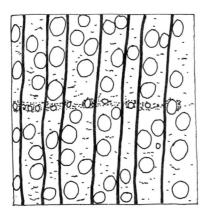

Gum canals are present and most of the vessels are solitary.

Laburnum
Laburnum anagyroides

Laburnum is one of the comparatively few timbers that show a ring-porous structure. It is the product of trees common to the north temperate regions but the wood is of little economic importance, though small quantities of it may be found on the market from time to time. It is moderately durable in exposed positions, works reasonably well in all operations, reacts satisfactorily to the normal types of decorative treatment and turns with good results. The timber is straight-grained but the texture is variable; it has an average dry weight of about 880 kg/m^3. Sapwood and heartwood are well marked, darkish olive brown in colour for the latter and yellowish for the thin sapwood, with a distinctive figuring but no characteristic taste or smell. Laburnum may be used for minor purposes such as inlaying, marquetry, turnery and small fancy articles. Complete cross-sections of small stems were inlaid in the past and referred to as 'oysters'.

The so-called Indian Laburnum is the wood of *Cassia fistula*. It is a strong and hard timber with a heartwood strongly resistant to termites and insects but its economic importance is purely local.

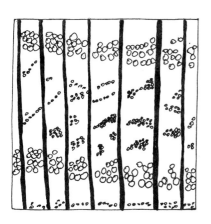

The wood has a ring-porous structure with latewood vessels either in tangential bands or loose clusters.

Lancewood

(Asta)

Oxandra lanceolata

Lancewood is a timber with an average weight of about 980 kg/m³ when dry, a straight grain and a fine and even texture. Only the sapwood is of any commercial importance, this being a darkish yellow in colour, whilst the heartwood is black, or almost black. Small quantities of the wood are often to be found on the general markets, though most of the trade is done with America, Jamaica and Tropical America which supply the timber, which is strong and non-durable. Lancewood splits easily and should be pre-bored for such operations as nailing and screwing but otherwise it creates few problems in working, whether under hand or machine processes. The timber turns satisfactorily and finishes well.

Typical uses include turnery, wheel spokes, parts of vehicles, shooting sticks, tool handles, billiard cues and other purposes for which a resilient timber is of value.

Burma Lancewood is a totally different reddish-brown timber which comes from *Homalium tomentosum* and is only used locally in India and Burma for construction and agricultural implements.

Lauan
Shorea, *Parashorea* and *Pentacme* species

The three genera listed are all important commercially and produce a variety of similar timbers; the light to medium weight timbers are named according to the country of origin. Lauan is the Philippine name for timbers of the three genera. Seraya is the Sabah name for timbers of *Shorea* and *Parashorea*. Meranti is the name for *Shorea* species grown in Malaysia, Sarawak, Brunei and Indonesia.

Each type can consist of a large number of species and the timbers are further divided into commercial categories on the basis of colour and weight; red, yellow and white groups are recognized in the timber trade and the red group may be available as light or dark or a mixture. Any group has several species in it and therefore is a little variable in properties.

Summary of Groups

1. Lightweight, pale red or pink timbers include:
   Light Red Meranti
   Light Red Seraya   Weight about 550 kg/m^3
   White Lauan

2. Medium weight, dark red timbers include:
   Dark Red Meranti
   Dark Red Seraya   Weight about 630 to 710 kg/m^3
   Red Lauan

3. Medium weight, yellow timbers include:
   Yellow Meranti
   Yellow Seraya   Weight about 660 kg/m^3

4. Medium weight, white or pale timbers include:
   White Meranti   Weight about 670 kg/m^3

(continued overleaf)

5. Heavyweight, red timbers include:
   Alan (Sarawak, Brunei)
   Red Selangan Batu (Sabah) } Weight about 850 to 880 kg/m³
   Red Balau (Malaysia)

6. Heavyweight, yellow or brown timbers include:
   Sal (India)
   Chan (Thailand)
   Balau (Malaysia) } Weight about 790 to 960 kg/m³
   Selangan Batu
   (Sarawak, Brunei, Sabah)

It can be seen from the list above that White Lauan and Red Lauan are not equivalents.

Red Lauan normally has a deeply interlocked grain that gives quarter-sawn stock a prominent striped figuring, while the heartwood is red or a dark red-brown in colour; the texture is coarse and uneven. The wood is sufficiently durable for all interior purposes and is in demand for such items as cabinet and furniture making, veneers and plywood. It works well but has a tendency to warp during seasoning.

White Lauan is usually provided by *Pentacme contorta*. It is not as durable as Red Lauan and contains numerous gum canals that may limit its usefulness. The surface is generally slightly lustrous. Texture is variable from reasonably fine to definitely coarse but seasoning and working qualities are good. White Lauan may be used for much the same purposes as Red.

White Seraya has a separate entry.

Laurel

Laurus nobilis

The Mediterranean regions are the natural home of the European species of this timber; the tree grows to a height of some 15 metres or more under favourable conditions. It is widely cultivated, being used principally as a shade tree, though its importance as a timber is negligible. The straight or occasionally interlocked grained wood has the typical mild fragrance of the Lauraceae family but has no distinctive taste. Laurel is whitish in colour, sometimes showing a light reddish tint and has a texture that may vary from fine to moderately fine and even. Sometimes known as Bay Laurel, the tree is also the source of a volatile oil that is an important article of commerce. The wood is suitable for small novelties, carvings, inlay and marquetry but is otherwise of little use. Drying and working qualities are quite satisfactory and it is only moderately heavy in weight. The degree of natural durability is not high.

Laurel, Chilean
Laurelia sempervirens

This timber is available only in small dimensions and is chiefly used for interior fittings, mouldings and cheap furniture, though it has some local importance for plywood manufacture. The wood is from a tree native to Central Chile.

The sapwood of the species is of a uniform greyish-brown colour, while the heartwood is of a purple-brown shade that is streaked with green or purple. The grain is almost invariably straight, the texture moderately fine and the average dry weight about 510 kg/m³. Planed surfaces of the wood are only slightly lustrous.

The shrinkage factors of Chilean Laurel are high and degrade from warping or checking may be high unless adequate precautions are taken; the timber is regarded as being below average in stability. It is of the class of timbers that are not naturally resistant to either fungal or insect attack; closely related species last well under water, with marked resistance to marine borers and Chilean Laurel will probably show the same qualities. Thick timber should be pre-bored for nails and screws and keen, thin-edged tools are essential in planing, though otherwise the wood creates few problems in machining.

Scalariform perforations are present.

Laurel, Indian
Terminalia species

(Taukkyan, Asna, Mutti)

The so-called Indian Laurel ranks as one of the most important of all the native Indian hardwoods. It is a timber of three species of *Terminalia* and has a fairly straight but variable grain, a coarse and uneven texture and an average weight of about 736 to 960 kg/m^3 for dry specimens. The sapwood and the heartwood of the species are clearly defined, the former being a reddish-white in colour, with the latter ranging from a light brown to a deep chocolate shade and showing darker, irregular markings. It is hard, dense and rather difficult to dry and is accordingly often girdled as a pre-felling treatment. Indian Laurel is naturally durable. The timber is somewhat difficult to work by hand but machines satisfactorily; nailing may be found difficult but glue adheres firmly and the timber also turns and polishes well. It may be used for such items as furniture manufacture, cabinet making, veneers, panelling, high quality joinery, tool handles and harbour piling. The wood structure closely resembles that of Afara.

Lignum Vitae
Guaiacum species

(Guayacan, Ironwood)

Three species from the West Indies and the Tropical American area constitute this timber.

Many generalizations such as 'the hardest timber in the world' have been made about Lignum Vitae but this and similar statements are not entirely accurate. The wood is however very hard and it is a difficult one to work whether by hand or machine. Lignum Vitae turns excellently and is naturally very resistant to fungal attack. The wood has a fine even texture and interlocked grain. A steady demand exists for the wood for certain specialized purposes and though it is expensive there has so far been little difficulty in fulfilling orders. It is a very heavy wood with an average dry weight of some 1150 to 1300 kg/m³, with a narrow yellowish or whitish sapwood and a heartwood dark greenish-brown to blackish in colour; the wood has a characteristic pleasant smell coupled with a bitter taste. Lignum Vitae is suitable for such items as pulley sheaves, stencil blocks, rollers and bowlers' 'woods'. The timber is so naturally oily that it has very marked properties of self-lubrication which make it especially suitable for some of the purposes mentioned above and also for wooden bearings for ships' propellor shafts.

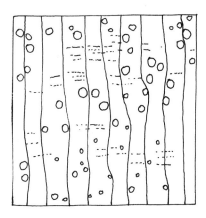

A slight indication of ring-porous structure sometimes shows. Ripple marks are present but very hard to see even with a lens. Lines of parenchyma are very faint.

Lime, European   (Linden)
Tilia vulgaris

Lime is a moderately soft and lightweight wood with a dry weight of about 540 kg/m³. It is a straight grained, fine and even textured wood; whitish or pale yellow to light brown in colour, with no distinctive taste or smell, no figuring and indistinguishable sapwood. In general appearance it closely resembles Poplar and Willow. The tree producing the timber is common in Russia, Germany and England and is a member of the same botanical family as Basswood.

Movement   – medium
Durability   – perishable
Treatability   – permeable
Workability   – works well and easily given sharp tools

Typical uses include carving, inlaying, marquetry work, cabinet making, small turned articles, paint brush handles and hatmakers' blocks.

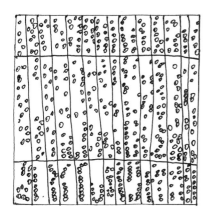

Lime is very similar to several other timbers and is best identified by microscopic features.

Locust, Honey

(Sweet Locust)

Gleditsia triacanthos

This timber is a native of the eastern areas of the United States. It has a wide sapwood, yellowish in colour and moderately well defined from the heart which varies from bright cherry-red to reddish-brown. Dried stock has an average weight of about 730 kg/m³; the grain is usually straight with the texture moderately open and somewhat uneven. Although the sapwood may be attacked by wood borer pests, the heartwood is strongly resistant to fungal infection.

Honey Locust is said to be moderately difficult or difficult to work and shows a tendency to split when being nailed or screwed. It responds well to finishing agents. The timber is used for furniture, structural work and general utility purposes.

Logwood
Haematoxylon campechianum

Logwood, which originates in the West Indies and Tropical America, has a dry weight of about 1000 kg/m³ The grain is interlocked and the texture coarse but fairly even; the heartwood and the sapwood are very sharply distinct, the former bright red and the latter whitish or yellowish. The wood has a distinctive sweet taste and a smell that resembles that of violets. Small quantities of the timber are imported in the form of slim logs but the economic importance of the tree rests on its usefulness in the dyeing industry rather than as a source of timber, though the wood is occasionally used for such secondary purposes as marquetry and inlaying; the wood also has certain medical uses. The timber is strong but brittle (needing pre-boring for nailing or screwing) and is durable for use in positions where it is exposed to the weather or rests in contact with the ground. It is rather difficult to work by hand but may be finished to a very smooth surface and takes a high polish. If chips of the wood are boiled in water the red dye is extracted and can be used to dye fabrics. The red colour in the timber fades to a reddish-brown on exposure to light.

Mahogany, African
Khaya species

There are five species of Khaya growing over East, Central and West Africa. *Khaya grandifoliola* and *Khaya senegalensis* produce timber weighing on average 740 kg/m^3 when dried and are best known as Heavy African Mahogany – this wood is darker as well as heavier.

Khaya ivorensis, *Khaya anthotheca* and *Khaya nyasica* form the bulk of imported African Mahogany and of these the first forms the major part. The name of this West African timber may be preceded by the name of the port or district from which the shipment was made, such as 'Lagos', etc. It has straight or interlocked grain, a moderately coarse texture and has a weight of anything from 400 to 800 kg/m^3 dried; the average is about 530 kg/m^3 and grading ensures that the extremes are avoided. The sapwood and the heartwood are not always well distinguished, the colour varying from light pinkish-brown to a deep reddish-brown with radially sawn timbers showing a striped or roe figuring.

Movement   –   small
Durability   –   moderately durable
Treatability   –   very resistant including the sapwood
Workability   –   generally good (except the heavier species) but quartered stock may be difficult to plane; finishes well

The original true Mahogany comes from Central and South America; African material is generally accepted now as a Mahogany (the two are close relatives in the same botanical family) except in North America where it is usually called Khaya Wood. Although rather inferior to American Mahogany, African Mahogany is widely used for furniture, cabinet making, high grade joinery, panelling, flooring and boat building. It does not bend well but is good for plywood and veneers. The two heavy species can be used for all the same purposes provided that the weight is no disadvantage. Brittleheart and tension wood are sometimes present.

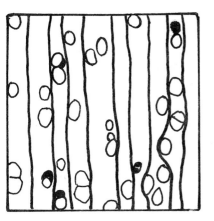

Growth rings are often not visible; parenchyma is also not to be seen; black gum deposits are common.

The word 'Mahogany' is much misused in the timber and furniture industries. The word should only be applied to species of *Swietenia* or *Khaya* and not to any timber which happens to have a similar colour but not necessarily the other properties.

The original Mahogany was the wood of *Swietenia mahagoni*; this was a beautiful dense wood weighing about 720 kg/m³ when dried and was used in such quantity that it became scarce and was replaced by a closely allied species *Swietenia macrophylla*. The original Mahogany can still be bought under the names Cuban or Spanish Mahogany but is too scarce and expensive for mass production work.

Swietenia macrophylla has been established for a long time as the normal wood of commerce – it is often called Honduras, Guatemalan, Mexican, etc. Mahogany according to its origin or port of shipment; it is normally well graded and can be expected to weigh about 560 kg/m³ when dried.

The colour varies from a light yellowish- or reddish-brown to a dark reddish-brown; the wood has a good natural lustre and tends to darken on exposure; the sapwood is yellowish-brown. The grain is mostly either straight or slightly interlocked but a great variety of mixtures of wave, stripe, mottle, blister etc. can be found and are frequently cut as veneers. The texture is moderately coarse and the timber steam-bends moderately well.

Movement   –   small
Durability   –   durable
Treatability   –   very resistant
Workability   –   easy to work and finishes well

Tension wood is sometimes present and will only finish well if very sharp cutters are used. The wood is used for high quality furniture and cabinet work, panelling, boat building, pattern making and veneers.

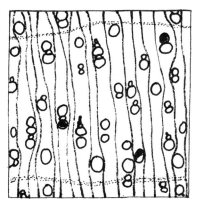

Vessels are usually smaller than in *Khaya* species. Growth rings show distinctly with a line of parenchyma. Dark gum is normally present in many vessels and often white deposits are present in vessels too. Ripple marks are often present.

Makoré (Baku, Douka)
Tieghemella heckelii

The sapwood and heartwood of this timber are well defined; the sapwood is pale yellow or whitish, while the heartwood is reddish-brown or purplish-brown and sometimes darkly streaked. The grain may be straight or slightly interlocked; often a mottle figure which is most attractive results from the irregular interlock. The texture is fine and even and the average dry weight is about 620 kg/m³. Planed surfaces of the wood are highly lustrous. The timber comes from various West African countries.

The timber has sometimes been described as Cherry Mahogany or African Cherry, though botanically it is not related to the true Cherry or Mahogany.

Movement   –   small
Durability   –   very durable
Treatability   –   very resistant including sapwood
Workability   –   rather difficult to work and blunts tools rapidly; finishes well; needs pre-boring

This wood contains silica, which causes the blunting effect. It also stains when in contact with iron in moist conditions and produces a fine dust in working which is irritant to many people. In spite of these facts the wood is well-liked because of its appearance. It is used for furniture and high quality joinery, threshold strips, flooring and vehicle building. Sliced veneers and plywood are also available.

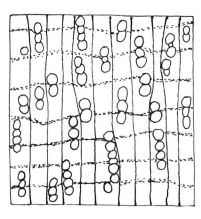

The radial groups of vessels are a striking feature of this timber.

Mandio

(Mandioqueira)

Qualea species

Three species occur in the north of South America and are marketed as a mixture.

The sapwood is whitish or yellowish and is clearly marked off from the rather featureless heartwood, which is usually of a golden reddish-brown shade, though occasional timbers show a purplish tinge. The planed surfaces of the wood are dull, the texture is moderately coarse but even and the grain is straight to interlocked. Average dry samples of the wood weigh about 600 kg/m^3. It is not an easy timber to work in most processes and needs care to bring it to a good surface. In general the wood responds well to finishing processes, though it does not polish well. The timber is classified as moderately durable and is reputed to respond reasonably well to treatment with preservative fluids. Mandio needs to be dried slowly to minimize warping that will be most noticeable in stock that has really deeply interlocked grain.

This wood is used for furniture, joinery and flooring.

Mansonia
Mansonia altissima

This species has a limited distribution in Nigeria, Ghana, the Ivory Coast and Cameroon. The sapwood is whitish and clearly distinct from the heartwood which is yellowish to greyish-brown, frequently with a lilac or purplish tone and some streaking. The fresh heartwood colour fades very rapidly on exposure and assumes a darkish straw colour. The grain is straight and it has a fine, even texture. The weight is about 600 kg/m³ dried.

Movement   –   medium
Durability   –   very durable
Treatability –   very resistant
Workability   –   easy to work and finishes well

This timber causes more complaints of irritant effects from the dust produced in working than any other; it affects eyes, nose, throat and chest and may also cause dermatitis. The effects quickly disappear but are best avoided and so very careful attention to dust extraction and use of masks is advisable.

Mansonia is used for furniture and joinery as an alternative to Walnut which it slightly resembles. It also finds some use for high quality cabinet work and television cabinets.

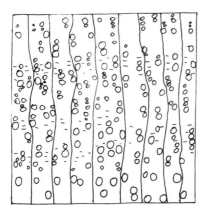

Fine lines of parenchyma can just be seen. Ripple marks are present.

Maple (various)
Acer species

Various species of maple are to be found in Europe, Asia and America but it is on the last continent that it achieves its greatest importance as a timber. The American species are dealt with below. In general the Old World species are pale brown in colour, showing a fairly distinctive growth ring figuring and have an average weight of between 660 and 690 kg/m³ when fully dry. The working qualities of the timber are satisfactory provided that the tools are kept keen and it will both turn and carve with good results. The grain of the timber is normally straight and the texture fine and even. Maple reacts well to almost every form of decorative treatment and in particular is capable of giving a good finish with wax polishes. The wood is perishable.

Furniture manufacture, cabinet work, flooring, musical instruments, panelling and wainscotting are among the uses for this timber.

Common Field Maple is the timber of *Acer campestre*. Other species include the Norway Maple (*Acer platanoides*) which is native to northern Europe but also to be found in America, the Japanese Maple (*Acer palmatum*), lighter both in colour and weight than the Common Field Maple and the Himalayan Maple which finds considerable use in India. All these species are similar in their general nature and properties but are rarely, if ever, marketed under these specific names with the possible exception of the Norway Maple.

Mineral streaks (greyish/brownish/greenish) may be present occasionally.

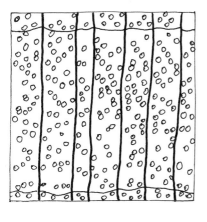

The terminal parenchyma marking the growth rings shows very clearly.

Maple, Queensland   (Australian Maple)
Flindersia brayleyana

This is one of a group of *Flindersia* species growing in Australia and Papua New Guinea and which are incorrectly and often misleadingly described as types of Maple or Ash.

Queensland Maple is a light brown or pinkish-brown wood with an interlocked or wavy grain and a medium coarse, even texture. It rather resembles a pale African Mahogany except that it is lighter in weight – about 560 kg/m^3 when dried.

The timber is prone to develop various forms of degrade during drying but machines reasonably easily.

It is used for furniture, cabinet making, high quality joinery, veneer and plywood.

Maple, Rock
Acer saccharum

(Hard, Sugar or White Maple)

Rock Maple may also be given the alternative name of just plain Maple. It has an average dry weight of about 720 kg/m³, the grain is either straight, wavy or curly and the texture is fine and even. The timber is a lightish brown in colour and has a distinctive growth ring figuring. It is one of the most popular furniture and cabinet woods of the American market and is the product of a tree to be found growing in Canada, Newfoundland and the northern United States. Pith flecks are infrequently present.

Movement   – medium
Durability   – non-durable
Treatability   – resistant
Workability   – rather difficult to work; needs pre-boring; finishes well

This species is sometimes attacked by an insect which causes the production of the well-known 'birds eye' figure; this is usually peeled into beautiful veneers. The solid wood is hard and strong and is used for flooring (domestic, ballroom, gymnasium or heavy industrial), sports goods, shoe lasts, musical instruments and plywood.

The structure is the same as for the European species. The tree is also the source of Maple sugar and Maple syrup. Other North American species include the Red Maple, also known as the Swamp or Soft Maple; the Black Maple (*Acer nigrum*), with a reddish-brown heartwood and a whitish sapwood but not as dense and hard as the Rock Maple; and the Striped Bark Maple (*Acer striatum*), of little importance as a timber. The Soft or Silver Maple (*Acer saccharinum*) is not as highly esteemed as the Rock Maple but provides a lighter weight reddish-brown timber that finds considerable use. The Philippine Maple is *Acer niveum*.

Maple, Rose
Cryptocarya erythroxylon

This timber has a range of weight from 640 to 800 kg/m^3 when fully dried. Generally the grain is straight but it may be slightly wavy and the texture is coarse and even. The heartwood of the timber ranges from pale brown to medium brown in shade, with orange or pink tints, quarter-sawn surfaces showing a ray figuring; the wood has no distinctive taste but has a rather characteristic fragrant smell. Rose Maple is a moderately durable timber for external use and is the product of a tree native to New South Wales and Queensland. It is used in Australia for such purposes as cabinet making, panelling and wainscotting but the wood is not well-known on the general market and it has no outstanding characteristics that are likely to result in a demand for supplies from places outside its native continent.

Marblewood, Andaman
Diospyros marmorata

Andaman Marblewood is a straight grained, fine and evenly textured wood that has an attractive figuring of dark or black bands on a grey-brown or light grey background. The timber is the product of a tree of the Andaman and Nicobar Islands and it is probable that it would be greatly esteemed as a decorative timber if it were more readily obtainable. At present the bulk of the timber goes to the neighbouring Indian markets. The tree is of the same genus as the true ebonies and Marblewood shows all the characteristic advantages and disadvantages of that timber. It is not easy to season and is difficult to work, though it is excellent for carving and turning. The wood is classified as durable in exposed positions, though this property is of little importance for the purposes for which it is customarily used.

Typical uses for the wood include carving, inlaying, turning, fine cabinet making and so on. Andaman Marblewood is sometimes given the name of Zebrawood, though this title belongs preferably to the timber of *Astronium fraxinifolium* from Brazil.

The structure of this wood is not distinguishable from the other ebonies with a hand lens although the colouring is distinctive.

Meranti (various)

Shorea species

The species grow in various South East Asian countries and are known as Meranti, Seraya or Lauan, depending upon their origin. Meranti is the commonest name in the United Kingdom and the type most used is the red group. Most of it is light red or pinkish in colour; the average dry weight is 550 kg/m^3 but it may be as low as 400 or as high as 705 – the dark red material is at the upper end of this range. Because of the mixing of colour and weight some care may be needed in selecting pieces to match each other. More detail of this group is given under the entry for Lauan.

Light Red Meranti is light red to pinkish-brown when freshly cut. The texture is rather coarse but even; the grain varies from straight to interlocked and stripe figure is often present. Gum canals are often present and may show as whitish lines on planed and sanded surfaces but the gum is white and crystalline, not sticky. The wood takes stains and finishes well.

Movement   –   small
Durability   –   moderate
Treatability –   resistant at best, including sapwood
Workability   –   no difficulties

This type of Meranti is used for joinery, panelling, louvre doors, furniture, light construction and plywood. The timber is sometimes confused with African Mahogany and may be called Philippine Mahogany because of this resemblance. The other types of Meranti (Dark Red, Yellow and White) all tend to be heavier and perhaps more suited to structural uses; but they may also be used for the same purposes as Light Red Meranti.

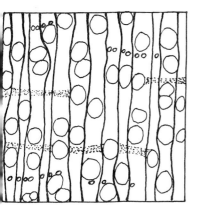

The white crystalline deposits in gum canals are usually quite easily seen with the naked eye.

Light Red Meranti

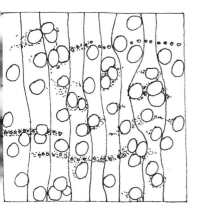

The fibrous background is darker and denser than in Light Red Meranti. Gum canals are again visible.

Dark Red Meranti

Mora

Mora excelsa

Mora is not well-known on the world markets, but it ranks as one of the most, if not the most, important timbers of British Guiana, although its uses are somewhat limited. It has an average weight of 1020 kg/m³ fully dried; the grain is interlocked and the texture coarse and uneven. The sapwood and the heartwood are well distinguished, the wide sapwood being yellowish and the heartwood reddish-brown or darker in shade. There is no typical fragrance to the wood but it has a bitter taste. Mora is naturally durable and the wood is hard and rather difficult to work by hand or machine. It is also rather difficult to dry and suffers splits, checks and distortion.

Typical uses for the timber include ship-building, paving blocks, sea defences and railway sleepers. Sufficient supplies of the wood are available to meet the present demand and there is little likelihood of that demand being increased, as it has little to recommend it over other timbers better known to the trade.

On the American market, Mora timber is sometimes known as Nato. A related and very similar species from the Guyanas (*Mora gonggrijpii*) is sold under the name of Morabukea.

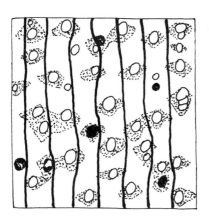

In some samples growth rings show faintly. Ripple marks may be present.

Mtambara

Cephalosphaera usambarensis

This species is mostly found in Tanzania. The sapwood is not easily distinguished from the pale pinkish-brown heartwood but the heartwood darkens more with age. The grain is straight and the texture is moderately fine and even. The dried weight is about 590 kg/m³. The wood is unexceptional in appearance.

Movement   –   medium
Durability   –   perishable
Treatability –   moderately resistant
Workability   –   very easy to work and finishes well

The timber is a useful utility species suitable for interior joinery, plywood and furniture construction.

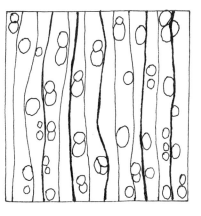

No well-marked structural features are evident.

Muhuhu
Brachylaena hutchinsii

This East African timber is virtually a special purpose species. It is a very hard and dense timber of unusual colour – when fresh the yellowish-greenish-brown is quite distinctive but it tends to darken slowly to a mid-brown shade. The texture is very fine and even and although it is difficult to work because of its hardness, a very good finish is obtainable. The weight is a little variable but averages at about 930 kg/m^3 dried. The hardness provides good resistance to abrasion and indentation and so the wood has found a particular use as a beautiful flooring timber suitable for any extent or type of traffic. It is used locally for carving and turnery also.

Mulberry
Morus species

Mulberry is the timber of various species of *Morus* growing in Europe, India and China and is an irregularly grained, coarse and uneven textured wood, yellow to red-brown in colour in the heartwood and with an average dry weight of about 640 kg/m³. It is a difficult wood to dry without degrade and it is durable. The timber is primarily a cabinet wood and is mildly decorative in that quarter-sawn stock has a well-marked silver figure. In most hand and machine operations the wood works well and it turns rather better than its uneven texture indicates; under steam treatments the wood bends well.

Although normally marketed together with other *Morus* species as Mulberry, the Red Mulberry (*Morus rubra*) may occasionally reach the market under its own name. Although this tree will grow in Europe it is really a native of the North American continent, reaching its best development in the lower regions of the Mississippi basin, where it may reach a height of some 24 metres. In general the properties of the wood conform with those of common Mulberry, though Red Mulberry is slightly the heavier of the two timbers, somewhat harder and consequently rather more difficult to work. It is reasonably strong for its weight (680 kg/m³) and is generally regarded as a cabinet wood. Reddish-brown in colour and with an irregular grain, Red Mulberry has a finer and more even texture than is found in common Mulberry. The Indian Mulberry is *Morus indica*; this species is used in sports goods.

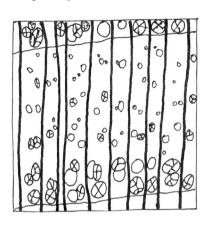

This is another of the very few ring-porous timbers. It is similar in structure to the Elms but the tangential arrangement of latewood vessels is not so marked in Mulberry.

Muninga   (Mninga, Ambila)

Pterocarpus angolensis

This timber may also be known locally as Kajat or Bloodwood and is used for cabinet work, interior fittings, furniture and domestic flooring. It is a close botanical relative of the true padauks.

Muninga has a whitish to pale yellow sapwood that is sharply defined from the heartwood which varies in colour from a golden brown to a reddish-brown background tint, handsomely figured with golden-yellow or reddish streaks. The average weight is about 580 kg/m^3 dried but is rather variable. The grain is mostly irregularly interlocked and the texture, while even, is rather coarse. Freshly planed timber has a silky lustre and often has a faint, pleasant perfume.

Movement	–	small
Durability	–	very durable
Treatability	–	resistant, including the sapwood
Workability	–	fairly easy to work except for interlocked grain; finishes well

Utilization of this wood can be complicated by the tendency of the logs to be either golden brown or red brown. Material needs to be carefully matched; it is also prone to show rather conspicuous white deposits which show through polish. They can be removed before polishing by rubbing the surface with white spirit and allowing it to dry.

Muninga is well worth taking trouble with because it provides a very handsome appearance for furniture and cabinet work.

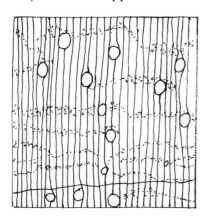

The structure is very similar to that of Andaman Padauk except that Padauk has narrower bands of parenchyma.

Myrtle
Myrtus communis

Although this is a member of the same botanical family as that producing the important *Eucalyptus* varieties, this timber is of little commercial importance. The tree producing the wood is of medium growth habit and is common to the Mediterranean regions and the more temperate parts of Asia; the bark of the tree is often used for tannin extraction. With drying treatments the lightish coloured wood is very refractory and it is difficult to handle in all hand or machine processes, largely because of its irregular grain. Timber of the species normally weighs about 800 kg/m³ when dried and the texture is moderately coarse and uneven. No special demand exists for the timber and no demand is likely to arise for it.

The Tasmanian Myrtle, also known as Myrtle Beech or Australian Myrtle, is the timber of *Nothofagus cunninghamii* with an average dry weight of 740 kg/m³, a straight or shallowly interlocked grain and a very fine and even texture. It is a timber of Tasmania and the southern and eastern regions of Victoria and is a non-durable wood that seems to be rather susceptible to sap-satin fungal infection. Tasmanian Myrtle works reasonably well, whether by hand or by machine, finishes to a clean surface, turns satisfactorily and reacts well to most types of wood finishing treatment. Typical uses for the wood include cabinet making, turnery, shoe heels, flooring, piano cases, etc. The sapwood is almost colourless and the heartwood pink or pinkish-brown. The wood is not likely to become an important factor on the world markets. Scalariform perforations are usually present.

Niangon   (Nyankom)
Tarrietia utilis

Niangon grows in most countries of West Africa. Although this timber is primarily used for such purposes as joinery and cabinet work, it is really in the nature of a general utility wood, for it may be used for external and structural work. To some extent the uses to which the wood will be put will depend on the amount of resin that the sample shows and which gives it a greasy feel. The heartwood is of a pale red-brown colour; quarter-sawn stock has a ribbon stripe figure caused by the interlocked grain and it also shows an attractive ray flecking. The texture is rather coarse and planed surfaces of the wood feel slightly greasy to the touch. The average weight is about 620 kg/m^3 for dried material. The more resinous logs may be found moderately difficult to work, causing gumming up of saw teeth and plane irons and trouble in polishing. Samples having a lower resin content will work quite satisfactorily. Brittleheart is a fairly common defect and stock needs careful selection before use for purposes where strength is of importance. Niangon tends to move freely if it is not very carefully dried; it is not a difficult timber to dry though it may tend to split slightly during the process. The trees grow to a good height so that long, knot-free timbers of this durable species can be obtained.

The wood looks not unlike a very coarse mahogany and can look quite good for furniture and joinery including outdoor purposes.

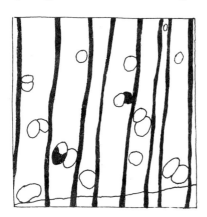

The rays are quite broad and easily seen. Fine lines of parenchyma are present but not easily seen.

Oak, American Red
Quercus species

(Northern, Southern, Spanish or
Swamp Red Oak)

Quite a large number of *Quercus* species grow in North America; they fall naturally into two botanical groups which the timber trade classifies as Red or White Oaks (European Oak is a white oak type).

American Red Oak is generally regarded as being inferior to European or American White Oak; it is somewhat heavier at 770 kg/m³ dried and it is also coarser, non-durable and resistant to preservative treatment although it is almost without tyloses. Its large rays are not so large and therefore the silver figure is not so marked. The colour usually shows a distinct pink or reddish tinge to the pale brown and it looks a little raw in consequence.

The precise quality will depend upon the area of origin and the grading but it is mostly used for less expensive furniture, flooring and interior joinery. Not suitable for cooperage or outdoor uses.

The individually distinct latewood vessels and lack of tyloses make this wood unlike the White Oak.

Oak, American White
(White, Chestnut or Overcup Oaks)

Quercus species

This timber also comes from the Eastern regions of Canada and the United States of America and, like the American Red Oak, is composed of several species sold as a mixture or individually.

The wood is a pale yellow brown to mid brown with a very distinct sapwood. The grain is straight and the texture fairly coarse. Because it is a ring-porous species it shows a prominent growth ring figure; it also has rays of two sizes like all true oaks and so shows a good silver figure on quartered surfaces. The grading of this timber is very important because when slowly grown it can be very weak. Like European Oak it shows staining when in contact with iron under moist conditions and it is acid enough to corrode many metals. The weight is variable but tends to be heavier than European Oak at about 750 kg/m³ when dried.

Movement   –   medium
Durability   –   durable
Treatability   –   very resistant including sapwood
Workability   –   tends to be easier to work than European Oak but this does depend on its weight; finishes well but needs pre-boring

It is suitable for all the same purposes as the European Oak – furniture, construction, joinery, cooperage, flooring, etc.

In structure it is likely to appear identical to European Oak.

Oak, Beef
Grevillea striata

This is an example of a mis-named timber, for Beef Oak has no affinity with the botanical family producing the true oaks. It is an Australian timber that is the product of a tree of New South Wales and Queensland and the wood has a purely local use for fencing posts, turnery, gates, fancy articles and general utility purposes. The heartwood ranges from red-brown to dark red-brown in colour, though the tint is uniform in any one specimen; the grain is irregular, the texture medium coarse and uneven and the average dry weight about 850 kg/m³. Beef Oak is moderately easy to work by hand, turns well and is reasonably resistant to the attack of wood-rotting fungi when used in exposed positions. The wood is not well-known outside the Australian continent and there is no likelihood of a greatly increased demand for it.

Other varieties of Australian Oak are provided by *Casuarina* species. She Oak (*Casuarina fraseriana*) is normally available only in small dimensions and it also suffers from the disadvantage of being rather difficult to work. The drying and the durability qualities are up to standard. Swamp Oak is good as regards drying but is not easy to work. It is the product of *Casuarina suberosa* and, like the preceding species, is of little commercial importance. Honeysuckle Oak (*Banksia integrifolia*) is also used for cabinet making and allied purposes. It may be pinkish to a definite red in colour, average dry specimens weighing 560 kg/m³. Working qualities of this wood are not always satisfactory and its durability is not good. There is little likelihood of its being exploited on a big commercial scale.

Oak, European

Quercus robur and *Quercus petraea*

Oak is the common name used to describe many hard and strong timbers, though the true oaks belong only to the genus *Quercus*. Some of the more important oaks are treated in separate entries below. The name Oak is used very loosely in much the same way as Mahogany is.

English Oak is the timber of *Quercus robor* and *Quercus petraea*. In weight it is about 720 kg/m³ when dried, the grain is generally straight and the texture coarse and uneven. The sapwood and the heartwood of the timber are not always sharply delimited, the sapwood being lightish in colour, while the heartwood may range from the same tint to a warm, rich brown. Flat-sawn timber has a very distinctive figuring due to conspicuous growth rings, while radially cut stock has a very pronounced silver figure. Of all the true and so-called oaks this timber enjoys the greatest popularity, though it is very variable in its quality and needs careful grading.

Movement   – medium
Durability   – durable
Treatability – very resistant
Workability   – ease of working depends very much on weight; finishes well but needs pre-boring

The rate at which it has grown will affect the texture, weight and strength of European Oak very much. Average material is reasonable to work but heavy stock is very difficult. The wood contains much tannin and this produces dark blue-black stains when in contact with iron under damp conditions. Oak is acidic and will attack many metals if they are not galvanized or painted.

Uses for the wood include: furniture, panelling, interior fittings, railway sleepers, coffins, cooperage, cabinet making, dowels, musical instruments, ship-building and vehicle-building, wainscotting and mouldings. Oak has a commercial use apart from its use as a timber in that the bark yields tannin in considerable quantities. It also steam-bends quite well and peels or slices into decorative veneers. The durability of Oak makes it popular for gates, fences, posts and thresholds.

The occasional very wide rays are quite easily seen and give rise to silver figure. The vessel rings contain many tyloses.

Oak

other true *Quercus* species

Japanese Oak is imported regularly into the United Kingdom, principally in the form of veneer and solids. The wood is coarser in texture than the European species and is normally lighter in colour. It is easier to work than the majority of true oaks and the silver grain figure is less prominent. Japanese Oak is not obtainable in big dimensions and is the timber of *Quercus grosseserrata*. Much of the early Japanese Oak imported into the United Kingdom was very slowly grown, coarse, lightweight and weak but it is now well graded and of good quality.

Indian White Oak is not rated as a commercially important wood and in its region of natural growth is often used only for fuel or charcoal. It is the product of *Quercus incana*, with a warm reddish-brown coloured heartwood.

Dyer's Oak is of some importance in that the bark yields a preparation in use in the dyeing and tanning of leather. It comes from *Quercus tinctoria* but is insignificant as a wood.

Baltic Oak is provided by the same species that constitute the common English Oak but is usually less strong and of smaller dimensions. It is sometimes classified by the title of its port of shipment as Memel, Danzig etc. Oak and in general its working and other qualities are the same as those of the English wood but in drying treatments it is often found to develop cup shakes. A considerable export trade exists with the timber.

There is also an Evergreen Oak (*Quercus ilex*) which is better known as Holm Oak. This one is unusual in that the growth rings are very inconspicuous – the wood is diffuse-porous and appears to have lost its rings of earlywood large vessels. It is heavier and more difficult in most operations and tends to be used for only for the rougher purposes that Oak is suitable for.

Oak, Satin
Embothrium wickhami

Satin Oak is very closely related to the popular Australian Silky Oak. Any fully dried specimen of the timber should weigh between 400 and 560 kg/m^3, have a straight grain and show a texture that is moderately coarse and uneven. The heartwood of the timber is pinkish in colour, quarter-sawn having a silver grain figuring reminiscent of the *Quercus* species and the Silky Oaks. Satin Oak is an easy timber to work, whether by hand or by machine and the dulling effect on tools is very slight. The wood is naturally resistant to decay when used in positions unprotected from the weather or when placed in contact with the soil.

It may be used for cabinet making, panelling, mouldings and similar purposes and it is also suitable for use as roofing shingles, though it is not a good structural timber. There is no likelihood of a big demand for the timber arising on the market apart from one that is purely local. A closely related species is Notro (*Embothrium coccineum*) from South America.

Oak, Tasmanian
Eucalyptus species

A timber closely related to Jarrah and therefore not a true oak or ash at all but nevertheless given such local names as Victorian or Australian Oak, or Alpine, Red or Mountain Ash. It is also described as Messmate Stringybark. Consignments are made up from three species of *Eucalyptus* (*Eucalyptus obliqua*, *Eucalyptus delagatensis* and *Eucalyptus regnans*).

The timber is commonly used for cabinet work, interior joinery, furniture, cooperage, flooring and plywood. It is moderately good wood for steam-bending, works well and responds satisfactorily to finishing agents. Some care is needed in drying.

The timber varies from light to deep brown in colour, with a poorly defined sapwood. It often has a prominent growth ring figure. The grain is variable but mostly straight, the texture moderately coarse but even and the average dried weight about 700 kg/m³. The wood quite often contains prominent dark gum veins.

Oak, Tulip
Argyrodendron trifoliatum

This timber has a weight that may vary between 800 and 900 kg/m³ for fully dried stock, a grain that may be either straight or shallowly interlocked and a texture that is variable from medium coarse to coarse but even. The sapwood and the heartwood are usually well distinguished; the sapwood is white to light brown in colour and the heartwood dark brown, with radially sawn timbers showing silver figure. The timber may be marketed under the names of Red, Brown or Blush Tulip Oak according to the predominating tone but there are no essential differences between them. The tree is not a member of the true oak family but timbers of large sizes are not uncommon.

Tulip Oak is rather hard to work but can be surfaced well; there is a tendency to brittleness and therefore the wood needs to be pre-bored for various operations. It is not naturally resistant to wood-rotting fungal attack but it seasons well.

The timber shows a few large rays; the silver figure that shows as a result is doubtless the reason for calling it an oak. The small rays are storied and therefore a ripple mark shows.

Obeche

Obeche (Wawa, Arere)
Triplochiton scleroxylon

Due to its light weight (380 kg/m^3 when dried) Obeche has become one of the most popular of the imported Africa species for core-stock, furniture parts, interior fittings and so on; it has also been found of considerable use for food containers on account of its non-tainting properties.

Lightness is not the only good feature of Obeche. Although soft and liable to crumble under dull tools, Obeche will work well both in hand and machine processes if proper equipment and sharp cutting edges are used. When possible, glue should be used in preference to screws or nails, as some samples are too soft to hold the screws satisfactorily. Some care is needed in filling because the texture is moderately coarse, though Obeche will stain evenly and also polish well. Sap-staining is a common defect which is very obvious because the wood is white to pale straw in colour. The grain is interlocked but figure shows very little because of the paleness of the wood.

Movement   – small
Durability   – non-durable
Treatability   – resistant
Workability   – easy to work and finishes well

The wood peels quite well and is available as plywood or in core plywoods.

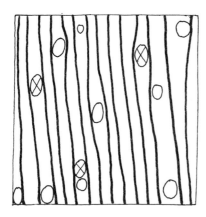

Fine lines of parenchyma are present but not easy to see. Tyloses are often present.

Odoko
Scottellia coriacea

The sapwood and the heartwood of this West African species are not clearly defined, the timber being a light yellowish shade and rather featureless apart from a prominent silver figure on quarter-sawn stock. Odoko is a straight or shallowly interlocked grained wood that has a fine and even texture, a mildly lustrous surface and an average dry weight of about 620 kg/m^3. It is obtainable is good dimensions. Odoko needs a certain amount of care in seasoning as it may develop splits.

Movement   – medium
Durability   – non-durable
Treatability – permeable
Workability   – fairly easy to work and finishes well; needs pre-boring

Odoko is a general utility timber for many domestic applications. Can be used for furniture, joinery and flooring.

Ogea

Daniellia ogea and *Daniellia thurifera*

These species are found in various West African countries. Other *Daniellia* species exist but should not be confused with them.

The sapwood is very wide (100–180 mm) and distinct from the pale reddish or golden brown heartwood which may show some streaks. The grain is usually slightly interlocked and the texture moderately coarse.

Movement   –   medium
Durability   –   perishable
Treatability   –   moderately resistant
Workability   –   generally easy to work but sometimes rather woolly

The wood is regarded as a very ordinary utility species. It is used for packaging, light joinery and core veneers for plywood.

The structure shows vertical gum canals and ripple marks. Brittle-heart is very often present in larger logs.

Okwen

Brachystegia species

Four species of *Brachystegia* grow in West Africa and tend to be mixed together. The weights are a little variable, ranging from 530 to 770 kg/m³ when dried – the ease of working varies with the weight. The heartwood colour is light to dark brown and may often show a fairly regular colour striping. The grain is often deeply interlocked and the texture medium to coarse.

Movement	–	medium
Durability	–	moderately durable
Treatability	–	very resistant
Workability	–	difficult by hand but reasonable with machine tools

The wood is as strong as Oak but does not finish so well, so its use has been limited. It is suitable for vehicle building and general construction where good durability is not essential.

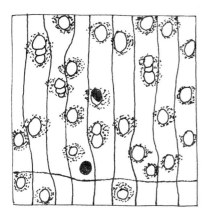

A very easily seen ripple mark is present; the terminal and aliform parenchyma is also most noticeable.

Olive, East African (Musheragi)

Olea hochstetteri

This species is pale to mid-brown and is attractively marked with brown, grey and black streaks in an irregular way; the sapwood is pale and without streaks. The texture is very fine and even and the grain slightly interlocked. The timber weighs between 830 and 1020 kg/m^3 when dried.

Movement	–	large
Durability	–	moderately durable
Treatability	–	moderately resistant
Workability	–	difficult to work but finishes well; requires pre-boring

The wood has good resistance to abrasion and makes an excellent domestic or heavy-duty flooring timber. It is also used for furniture, panelling and turnery.

The structure is very similar to that of European Olive except that the vessels are in radial groups of about 3 to 7.

Olive, European
Olea europaea

This particular species of olive grows in the Mediterranean regions of Europe, being a low tree of spreading growth habit, long life and slow growth. Economically it is more important for the olives and oil it produces than it is for its timber, though the fine-textured wood is used for carving, turnery, small fancy articles, novelties and similar small items. The timber is not naturally durable for use in unprotected positions but does not create any undue problems in drying. It works well in all hand and machine processes, does not blunt tools unduly and is an exceptionally good carving and turnery wood on account of the firmness and closeness of the texture. The grain of the timber may be either straight or shallowly interlocked, the average dry weight being from 800 to 900 kg/m³; the heartwood of the species is lightish brown in colour usually with darker brown stripes and streaks and sometimes showing a greenish tinge.

The structure of the wood is so fine that it is difficult to see even with a lens. The vessels are mostly single but sometimes in short radial groups of two or three.

Olive, Indian
Olea ferruginea

This is the Indian species of Olive and though the wood is of considerable importance in India it is not important on the world markets. The wood is naturally resistant to fungal attack and works well under most tool operations, though it may sometimes show a tendency to pick up on the machine planer. Fine seasoning checks are frequently found in the converted timber of the species, although the wood is not difficult to handle under drying. Local uses for Indian Olive include the making of tool handles, veneers, panelling, cabinet making, walking sticks, inlay work and turnery. The timber has an average dry weight of about 1000 kg/m³, the grain may be either straight or shallowly interlocked and the texture is normally very fine and even. The sapwood and the heartwood are usually very sharply distinguished; the sapwood is whitish in colour while the heartwood is light brown to deep purple, sometimes figured with darker bands of colour. The structure is similar to that of European Olive.

Opepe

Nauclea diderrichii

This West African timber has gained considerable popularity for such purposes as decorative heavy-duty flooring, piling and dock work and general construction. The wood has an interlocked grain which gives an attractive stripe figure on quarter-sawn stock. The heartwood is of a uniform mustard yellowish colour. The texture is moderately coarse and the average weight of dried stock is 740 kg/m³.

Movement   –   small
Durability   –   very durable
Treatability   –   moderately resistant
Workability   –   moderately easy except on the quarter; finishes well but needs pre-boring

The wood is available in large sizes and because of its strength and durability is well suited for construction and outdoor work. It is popular for harbours and sea defences. It is also suitable for domestic flooring, railway sleepers and exterior joinery.

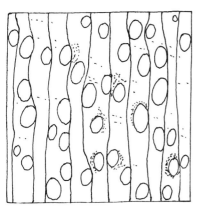

The vessels are distinctly oval in shape and are nearly always solitary.

Padauk, Andaman

(Padauk)

Pterocarpus dalbergioides

Andaman Padauk comes from the Andaman Islands and has an inter-locked or irregular grain and a texture that is variable but normally coarse; average timbers of the species weigh about 770 kg/m^3 when dried. The sapwood and the heartwood are sharply defined; the sap-wood is light in colour, while the heartwood may range from red-brown to warm red in shade and may have red or black streaks. Andaman Padauk is not difficult to dry either in air or in a kiln provided the process is slow. It is not easy to work, has a definite dulling effect on tool edges and is rather difficult to bring to a good surface, though it turns well and cuts satisfactorily for veneers, which are often beautifully patterned.

Movement   –   no information
Durability   –   very durable
Treatability   –   moderately resistant
Workability   –   not too easy, especially quartered stock; needs pre-boring and filling

It is a very hard and strong wood of considerable importance and may be used for such items as turnery, veneering, piano cases, balus-trades, high grade interior fittings, furniture, parquet and panelling.

Ripple marks are present. The various species of *Pterocarpus* are difficult to distinguish from one another.

Padauk, Burma

(Pradoo)

Pterocarpus macrocarpus

This timber shows a less wide range of colour and is not as ornamental as Andaman Padauk. It is naturally durable for use in positions unprotected from the weather and is especially resistant to termite attack; it is said to be difficult to impregnate with wood preservatives. The growing tree is often 'girdled' as a pre-felling treatment but during the drying operation surface checks are very likely to develop in the timber. The wood is rather harder to work than is Andaman Padauk and, being brittle, it should be pre-bored before nailing or screwing. Burma Padauk turns well and provided the grain is properly filled it is capable of taking a high polish. It is heavier than Andaman Padauk, averaging 850 kg/m^3 when dried; the grain is more deeply interlocked and the texture is moderately coarse. The heartwood of the timber is yellowish to darkish-red when it is first felled but later the wood tones down and becomes a uniform golden brown; quarter-sawn material has a narrow stripe figure. Structural characteristics are almost identical to those of Andaman Padauk.

Peach
Persica vulgaris

This timber comes from a slender tree of small growth habit, of more economic importance for its fruits than its timber, though that is suitable for certain minor uses such as inlaying and marquetry. It is closely related to the Almond, Plum and Cherry and consequently its timber resembles that obtained from those trees. The heartwood of the timber is reddish in shade and has an average weight of about 700 kg/m^3 when dry. The grain is somewhat variable but it is fine and even in texture. Quarter-sawn stock has a distinct silver figure and an attractive fragrance is also characteristic of the wood. Normal specimens of the timber are classed as difficult to work in all tool operations but it will surface well and its reactions to such treatments as staining and polishing are satisfactory. It soon succumbs to decay if it is used under damp conditions though it is satisfactory where it is protected against the weather. No specific demand exists for the timber, nor is such a demand likely to arise. It is most likely to be found in very old furniture.

Pear
Pyrus communis

The timber from this well-known tree only reaches the market in small quantities, the tree being native to both Europe and certain parts of Asia. It has an average dry weight of about 700 kg/m³, an irregular grain but a fine and even texture. Heartwood and sapwood are not sharply distinguished, the wood being a light reddish-brown in colour with no distinctive taste or fragrance. Pear is not a durable wood for use in unprotected positions. It works well under most hand or machine processes and may be given an excellent polish. The wood is only of limited use but it may be employed for the making of certain types of musical instruments and also finds use for inlaying, small fancy articles and ornamental turnery. There is only a small demand for the wood; it has been used in the past for furniture together with Apple etc. under the name of Fruitwood.

Peroba, White
Paratecoma peroba

The timber described here is from Eastern Brazil; the tree attains a good height but has a comparatively slender bole. Of rather less commercial importance is the Red Peroba, provided by various species of *Aspidosperma*.

White Peroba is essentially a timber for high-quality furniture, fittings, panelling and veneers but due to its high resistance to acids it has also been used for special purposes where this feature is of importance. The moderately lustrous heartwood of the timber is variable in colour from a pale olive to a dark brown but often shows a greenish, yellowish or reddish tinge; it is clearly defined from the much lighter sapwood. The timber has a grain that is variable from straight to interlocked, with a moderately fine and even texture. Fully dry samples of the wood weigh in the region of 750 kg/m³.

The working properties of White Peroba are very good in all processes, though it is sufficiently abrasive to dull cutting edges of tools more rapidly than the average wood. It responds satisfactorily to all the normal types of finishing agent. The strength properties are also good. White Peroba is very durable; its dust can be irritant to some machinists.

Apart from its furniture uses this timber is in demand in the chemical and food industries for the manufacture of vats and tanks. It has also been found suitable for flooring and boat construction.

Yellow vessel deposits are sometimes present as well as tyloses; ripple marks are often, but not always present.

Red Peroba (or Peroba Rosa) is a red-brown timber often streaked with purple or darker brown when fresh but fading on exposure. It is similar in most respects to White Peroba but is not suitable for chemical or foodstuff vats and is generally less well-known.

Plane, American

(Sycamore, Buttonwood)

Platanus occidentalis

This timber is known by the names Sycamore or Buttonwood in the United States of America.

It is the North American equivalent of the European Plane and is very like the Sycamore too. It is the product of the largest deciduous tree of the United States of America; the tree is of a rather rapid growth habit. When exposed to the weather or in contact with the soil the timber is not particularly durable but it is hard, tough and strong and finds a variety of uses. Although not a refractory timber under drying treatments it has a rather marked tendency to warp, more especially in thin stock. It is somewhat difficult to work, whether by hand or by machine but it finishes with a good surface and responds satisfactorily to decorative treatments such as staining, etc. The wood is light brown or yellowish in colour, with a very irregular grain but a fine and even texture; its average dry weight is about 560 kg/m^3.

The structure of the wood is very similar to that of European Plane. Scalariform perforations are present.

This species grows on a small scale in various parts of Europe. The wood has quite large rays and they are very obvious on a quartered surface – the pattern this produces has led to the name of Lacewood for sliced veneers. The name Plane is also applied to Sycamore (*Acer pseudoplatanus*) in Scotland and Northern England which can be confusing.

This timber, sometimes called London Plane in England, is useful not only for furniture manufacture and cabinet making but also as a structural wood. It is the product of a tree with a distinctive mottled bark that may grow to a height of some 25 metres. The timber is hard and strong but not difficult to work; it can be finished to a good clean surface. Plane is not naturally durable for use in positions unprotected from the weather or in contact with the ground. A certain amount of degrade may occur under drying processes as it is very apt to warp. The grain is normally straight and the texture fine and even, while the heartwood is either yellowish-white or brownish in colour. The weight of an average dry specimen will be found to be about 620 kg/m^3.

Another Plane species, though only of local importance, is *Platanus chiapensis* which produces the Mexican Plane or Alamo. It resembles its better-known counterpart very closely. The so-called Cape Plane is also a hardwood, the timber of *Ochnea arborea*, and although a good resilient timber it is only of local interest.

Most of the wood is used in ways which make the most of its decorative value – that is showing the Lacewood effect on quartered faces.

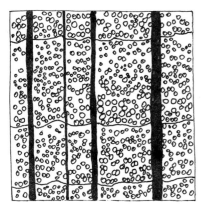

The wide rays are clearly visible without a lens. Scalariform perforations are present.

Plum
Prunus domestica

The tree producing this timber is common enough but is valued more for its fruits than for its timber, though the wood has a certain amount of use in the making of small fancy articles, novelties, furniture parts, inlaying and fine cabinet work.It is a timber that weighs about 640 kg/m^3 when dried, is straight-grained and has a moderately fine and even texture. The sapwood of the timber is normally broad while the heartwood is a deep brownish-red in colour. The drying qualities of the timber are quite satisfactory and it works well under tools in all operations, providing an exceptionally good timber both for carving and for engraving. The demand for the wood is practically non-existent.

Another timber known as Plum is the Sapodilla Plum. This hardwood timber is not of the same botanical family as that producing the true Plum; the tree (*Manilkara zapota*) flourishes in tropical America and is one of the sources of 'chicle' gum used for manufacture of chewing gum, while the bark of the tree also has certain medicinal properties. The greenish-coloured heartwood of the species is rated as very durable for use in exposed positions and small quantities of it may reach the market under the description of Black Bully. The timber is of little commercial importance.

Poplar (various)

Populus species

There are six principal species growing in Europe:

Populus canescens	– Grey Poplar
Populus nigra	– Black Poplar
Populus tremula	– European Aspen
Populus alba	– White Poplar
Populus italica	– Lombardy Poplar
Populus canadensis var *serotina*	– Black Italian Poplar
Populus robusta	– Black Poplar

The Black Poplars are most important as timber but all are very similar. The wood is soft (about 450 kg/m³ dried), fine textured, straight grained, without figure and almost without colour; a slight greyish or brownish tint may be present.

The quality depends very much on the origin and mixture of species – in general, better material comes from more northerly regions. The wood is strong and tough for its weight.

Movement	–	medium
Durability	–	perishable
Treatability	–	moderately resistant at best
Workability	–	depends on species but is likely to be difficult owing to woolliness unless very sharp tools are used.

Its woolliness makes it suitable for rough uses such as lorry and wagon bottoms. Peels well for veneers for vegetable and fruit box weaving and for match splints. Also used for low quality furniture parts, toys and hat blocks.

(continued overleaf)

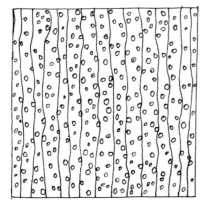

Identification is difficult because it resembles many other woods such as Willow.

Poplar, Canadian

(various)

Populus species

The name Canadian Poplar is used in the United Kingdom; Poplar in Canada is the name for a quite different timber (*Liriodendron tulipifera*). In Canada Poplars tend to be called cottonwoods or aspens, depending on the species involved, of which there are several.

Canadian Poplar is the timber of a small- to medium-sized tree that may reach an age of 200 years. It is widely distributed in Canada and the northern United States of America. It is a straight grained, fine and even-textured timber, lightish or pale brown in colour and having an average dry weight of about 450 kg/m³. The timber is non-durable if used in unprotected positions and cannot be easily treated with wood preservatives. Under drying treatments the wood may warp or twist considerably, resulting in a high degree of degrade, but it works easily under either hand or machine processes and the dulling effect on tools is not great; it carves and turns well and will take stain or other types of finishing agent satisfactorily. Canadian Poplar is not a very strong timber and is not suited for structural work but it is a good general utility wood. In general the wood closely resembles the European species and is used for much the same purposes.

Prima Vera

Tabebuia donnell-smithii

This Central American species is a yellowish-white with a slight pink tinge and has been wrongly known and sold as White Mahogany. The grain is irregularly interlocked and the texture medium to coarse. The broken stripe figure often provides an attractive mottle effect. The weight is about 450 kg/m^3 when dried. The wood is not difficult to work, except for planing, and finishes well. Pre-boring is required and some filling may be necessary for pinhole borer damage. Prima Vera is used for furniture, panelling, high quality joinery and veneers.

Purpleheart (Amaranth)

Peltogyne species

The many species providing Purpleheart occur widely throughout the Tropical American area and all have similar properties.

It is a heavy timber, dry specimens of which weigh about 1000 kg/m³. The grain may be either straight or wavy but the texture is almost invariably moderately fine and even. Sapwood and heartwood of the timber are well distinguished, the sapwood either whitish or streaked with light brown while the heartwood has a uniform purple-brown colour. The timber has quite a strong purple colour on first cutting it but this rapidly fades on exposure and may be very hard to detect on old surfaces. The purple will reappear if a thin shaving is cut to expose an inner surface.

The wood reaches the world markets in small quantities of small dimensions, though the bulk of the available supply goes to the American trade. It is a moderately difficult timber to work in any process but it turns well and can be finished to a good clean surface, although it has a severe blunting effect on tools. Purpleheart is very durable. Its outstanding quality is its ability to withstand sudden shocks, hence the wood is particularly suitable for such work as vehicle construction and the making of spokes and tool handles. It is also used for general construction both indoors and out. Its unusual colouring and stability make it popular for furniture, turnery, inlaying, billiard cue butts, billiard tables and flooring.

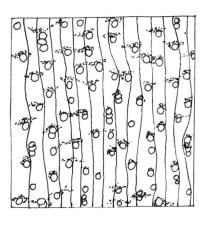

The parenchyma is a mixture of vasicentric and aliform. Ripple marks are present.

Pyinkado
Xylia dolabriformis

Pyinkado is one of the most useful of the structural woods of India and Burma and is obtainable in timbers of very good dimensions. It is a heavy timber, averaging when fully dry about 980 kg/m³. Pyinkado has a mild, non-distinctive smell, a grain that is shallowly interlocked and a moderately fine but even texture. The planed surfaces of the wood have an oily appearance. When freshly sawn the heartwood colour is of a dull red-brown shade but this darkens after exposure to air and the heartwood is then readily distinguishable from the thin, brownish-coloured sapwood. Darker streaks from fibrous zones may show in the heartwood.

Movement   – medium
Durability   – very durable
Treatability – very resistant including sapwood
Workability   – difficult to work and very blunting to tools; finishes
                 well with care but needs pre-boring

Typical uses for the wood include dock work, heavy duty flooring, railway sleepers, piling and telegraph poles. It is also a good domestic flooring timber and compares favourably with Maple, especially when used as plain-sawn material.

Pyinma (Jarul)

Lagerstroemia speciosa

Pyinma is a timber of Burma and adjacent parts of India that is not well known on the world markets but which is thought well of locally. The wood has an average dry weight of about 640 kg/m³, is straight-grained with the texture medium coarse but even; the heart-wood of the species is a pale red-brown in colour. Pyinma is a general utility timber that may be used for turnery, though selected material is suitable for panelling and similar purposes. The timber is moderately durable when exposed to the onset of wood-rotting fungi and is difficult to treat with preservatives, even by pressure processes. Pyinma works reasonably well under tools in all hand or machine processes and responds well to most types of finishing treatment.

Pyinma is a good timber for joinery, furniture, boat-building, turnery and light construction.

The structure of this timber shows a marked tendency towards the development of ring-porosity with rather oval vessels. Many tyloses are present.

A closely related species is the Indian Benteak (*Lagerstroemia lanceolata*); another is *Lagerstroemia hypoleuca* which is known as Andaman Pyinma – this is slightly heavier and better in many respects but seems to be used only locally.

Quaruba (Yemeri)
Vochysia species

Several species of *Vochysia* grow in Central and Tropical South America and it is at its best and most abundant in Brazil.

Quaruba is a timber of somewhat variable quality that may be used for interior joinery and fittings and most general purposes, though it does not appear to be greatly favoured for exterior work. The sapwood is of greater than average width and is not clearly marked off from the featureless pink or pale pinkish-brown coloured heartwood. The grain is variable, from straight to interlocked. The planed surfaces of the wood are moderately silky, while the texture is moderately coarse and even. Quaruba is not a particularly heavy species and when fully dry it has an average weight of about 500 kg/m^3. During drying quite a lot of distortion may occur. The durability is variable between the species; working is generally easy but a poor, rather fibrous surface results which needs a lot of sanding.

A number of factors operate against this timber. It is best used for light constructional work where it is out of sight or can be painted.

Quebracho
Schinopsis lorentzii

Quebracho is essentially a structural wood, being used for bridge building, railway sleepers, heavy construction and similar purposes. The wood is found in Brazil, Argentina and other South American countries; while timbers of good size are available, the quantity is rather restricted as the tree has considerable local importance as a source of tannin.

The sapwood and the heartwood of the species are well defined; the sapwood is grey or yellowish in colour while the freshly-cut heartwood is light red but this soon darkens to a brick-red shade – occasional samples of the wood may show blackish streaks. Planed surfaces of the wood are mildly silky, it has a bitter taste, the texture is fine and even and the grain is irregular and sometimes roey. Quebracho is a heavy species and when fully dry may weigh as much as 1150 kg/m^3.

The timber is very durable both to fungal and insect attack, though a certain amount of heart rot may be found in timber from over-mature trees. Checking and warping occurs to a considerable extent during seasoning, the trouble being particularly marked in timber of thin dimensions. After drying the wood is reasonably stable, while the strength properties are what might be expected of a timber of such high density. This is not an easy timber to work and has a tendency to chip out in drilling and moulding. In view of its likely use the finishing properties are comparatively unimportant but Quebracho can be given a very satisfactory polish.

Ramin   (Melawis)
Gonystylus macrophyllum

Ramin originates in Sarawak and Melawis in Malaysia.

The sapwood and the heartwood of the species are not easily distinguished for the timber is of a whitish to a pale yellow in colour and rather featureless. The grain may be either straight or shallowly interlocked, while the texture is moderately fine and even. Weight for dried stock is about 660 kg/m³. When green the timber has a strong and rather unpleasant odour and though this fades as the timber dries out it may return if the wood is rewetted. The seasoning properties are quite good though the timber shows a slight tendency to split and to develop minute surface cracks. If a suitable heat/humidity schedule is used the timber can be dried in a kiln with good results and after drying, whether by air or by kiln, the wood keeps its shape well. Some people suffer skin irritation when handling green timber; this is due to remaining sharp bark fibres – the dried wood causes no problems.

Movement   –   large
Durability   –   perishable
Treatability   –   permeable
Workability   –   easy to work; finishes well but needs pre-boring.

Ramin is a timber that is useful for general utility purposes whether it be furniture, interior joinery, toys, brush handles, plywood or a large variety of mouldings. Brittleheart is occasionally present.

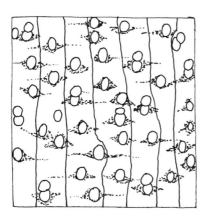

The aliform parenchyma is very striking – it tends to show long thin tangential points rather than short or medium ones.

Raspberry Jam
Acacia acuminata

This is a timber of Western Australia known by name only to the majority of timber users and is one that has a purely local importance, though even this is not particularly great. It has a heartwood that is dark brown to a darkish red-brown in colour, while quarter-sawn timber may show a striped figure owing to the grain, which is normally interlocked. The texture of the wood is fine and even and it has a distinctive smell that has been responsible for its name; the average weight of dried timber of the species is about 1000 kg/m³. Raspberry Jam is rather a difficult timber to work and is therefore frequently used in the form of round barked poles. It is hard and durable but has few, if any, outstanding properties that are likely to bring it into popular demand. The wood may be used for fencing posts and for general external work. There is little essential difference between this and other Australian species of *Acacia*, of which Australian Blackwood may be taken as an example.

Rauli (Chilean Beech)

Nothofagus procera

This timber is of the same botanical family as true Beech (*Fagus sylvatica*); species of *Fagus* do not grow in the Southern hemisphere – their equivalent in South America and Australasia is a range of species of *Nothofagus*, of which Rauli is perhaps the most well known. The timber comes from a tree of tall growth habit native to southern areas of Chile and timbers of good size are readily available.

The sapwood and heartwood of Rauli are not well defined, the timber being a rather featureless reddish-brown in colour; the timber does not show any ray flecking when cut on the quarter. Average weight is about 540 kg/m³ when fully dry, the grain is normally straight and the texture fine and even. The planed surfaces of the wood are non-lustrous.

Movement   –   small
Durability   –   durable
Treatability   –   moderately resistant
Workability   –   easy to work and finishes well

Rauli has been widely used as a substitute for European Beech but allowance must be made for its somewhat lower strength properties. It is suitable for joinery, furniture, flooring and has also been used satisfactorily for pattern making and vehicle construction.

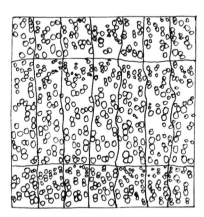

The structure is very similar to European Beech except that there are no large rays and therefore no flecks visible on quartered faces. Scalariform perforations are usually present in all species.

A closely related species, macroscopically almost indistinguishable from Rauli, is Coigue (*Nothofagus dombeyi*) which also grows in Chile. Coigue is rather paler in colour, heavier, slightly coarser in texture and subject to rather a lot of degrade in drying. It is generally regarded as being inferior to Rauli and is used mostly for rough work such as crates and boxes.

Another related species is *Nothofagus moorei*, the Negro-head Beech, sometimes known as Antarctic Beech. This is generally similar to Rauli but is heavier at 710 kg/m^3 when dried; it is mostly used in its native Australia for flooring and cabinet work.

Silver Beech, also known as Southland Beech, is the timber of *Nothofagus menziesii* from New Zealand. This again is very similar to Rauli and is used for much the same purposes.

Robinia
Robinia pseudoacacia

(False Acacia, Locust)

The tree providing this timber is normally regarded as only a decorative species but it produces a wood which, when available in quantity, is admirable for posts, gates and other exposed purposes, wagon bottoms and boat planking. The timber may be described in the United Kingdom as False Acacia but it is better known on the North American markets under the names of Locust or Black Locust. It grows in North America and has been introduced to Europe, Asia, North Africa and New Zealand.

Apart from some liability for powder-post beetle damage to occur in the sapwood, Robinia is very durable and the timber may be safely used for all external purposes without a preservative; it is extremely resistant to preservative treatment. Pre-boring for nails or screws is advisable but in all working processes it is satisfactory; it can be recommended for turnery purposes and will accept all normal finishing agents. Some care is needed during drying to avoid degrade from warping and shrinkage.

When freshly sawn the heartwood of Robinia is of a greenish shade but it soon darkens to an attractive golden-brown; the sapwood is narrow, yellowish and not very clearly defined. Planed surfaces of the wood are silky and when flat-sawn it has a prominent growth ring figure. The texture is coarse and uneven while the grain is usually straight. The weight is about 720 kg/m³ when dried.

The structure is ring-porous with numerous tyloses present. The latewood vessels are often in small clusters or in short tangential bands.

Rosewood, Brazilian

Dalbergia nigra

(Rio or Bahia Rosewood)

This timber has been steadily marketed over a period of centuries and supplies are now far less readily available than in the past. The wood is the product of a tall but slender tree of Tropical America, producing long but relatively narrow boards. Although chiefly used under cover the timber is naturally very durable; it works with some difficulty and because it may have a rather high oil content the wood will not always polish satisfactorily. The average run of dry timber weighs about 850 kg/m³; the grain is normally straight but in some specimens may be wavy and the texture is almost invariably medium coarse. Sapwood and heartwood are very sharply distinguished, the sapwood being almost colourless or lightish, with a heartwood ranging from shades of orange-brown through chocolate to a very deep violet brown; the colouring is often streaky. The timber has a distinctive mild fragrance and a faint taste. Typical uses for Brazilian Rosewood (which may also be called Jacaranda, though this is more in the nature of a local name) include cabinet making, panelling, furniture, piano cases and handles for certain types of cutlery, tools and instruments. Brazilian Rosewood may be summarized as being a first-class, highly important decorative cabinet timber.

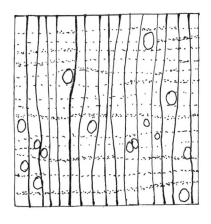

The lines of parenchyma are usually reasonably easy to see with a lens. A few chips of wood shaken with a little white spirit will give an orange/brown solution. Ripple marks may be present.

Rosewood, Honduras
Dalbergia stevensonii

Honduras Rosewood is heavy in weight, averaging about 960 kg/m^3 when dry; it normally has a straight grain and the texture is even but rather coarse. Sapwood and heartwood are sharply delimited: the sapwood is greyish in colour and the heartwood ranges from pinkish-brown to almost purple, with a figuring of darker lines or streaks; the wood has no taste but has a mild and distinctive smell. It is a very durable timber coming from a tree that reaches its best development in the coastal region of British Honduras. Under drying treatments it may prove somewhat refractory though the wood shows a high degree of stability once the drying process has been completed. It is not a very easy timber to work but it surfaces well; it turns well but very oily specimens create difficulty in polishing. Honduras Rosewood may be summed up as a slightly ornamental cabinet timber of more than local importance. The timber is closely related to Brazilian Rosewood and therefore their properties are very similar. It is one of the heavier Rosewoods and is used for cutlery handles and musical instruments.

Rosewood, Indian　　　　　　　(East Indian Rosewood)
Dalbergia latifolia

The timber comes from a large tree of slow growth habit that is widely distributed over India. It gives little trouble during slow drying and selected stock is one of the most attractive woods that can be used for furniture of all types and cabinet making. It is a naturally very durable timber when used in unprotected positions and has earned a good reputation for durability under water, though these properties are not of importance in view of the purpose for which it is used. Indian Rosewood is not an easy timber to work under hand or machine tools but it turns and carves well and may be brought to a good smooth surface; it responds well to finishing treatments and also polishes satisfactorily as its oil content is not as high as that of the other commercial rosewoods. It normally weighs about 850 kg/m³ when dry. In average timbers the sapwood is yellowish in colour and the heartwood medium brown to almost purple, with a darker figuring; fresh sawn timbers have the characteristic faint rose-like fragrance of the species. Indian Rosewood may also be sold under the title of Bombay Blackwood. It is often used in veneer form for furniture and is also used for certain musical instruments.

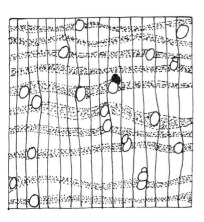

In general there are more vessels in Indian Rosewood than in Rio Rosewood. A few chips of Indian Rosewood shaken in white spirit give a purple/mauve solution. Ripple marks are present.

Sabicu (Jigue)

Lysiloma latisiliqua

The sapwood of this species is pale brown in colour and is not well defined from the golden-yellow or pale reddish-brown heartwood. The grain is apt to be variable, the texture is moderately fine and even, while the planed surfaces of the wood are lustrous. Sabicu is moderately heavy, weighing about 760 kg/m³ in the dry condition. It is from the West Indies and Florida and is sometimes sold as Jigue or Wild Tamarind. Very highly figured logs may sometimes be obtained but these are usually of small size. Sabicu may be used for furniture, interior fittings and, when available in sufficiently large dimensions, for structural work and ship-building. It was extensively used by Sheraton and other early cabinet makers.

The timber dries slowly but with only slight degrade; after drying it is a stable wood. The timber lasts well for external work even though untreated and it is also said to be difficult to impregnate satisfactorily with preservative fluids. Sabicu responds well to all normal types of finishing treatment and works satisfactorily both in hand and machine tool processes. The strength properties are good.

Sandalwood
Santalum album

The tree producing this timber is a native of the East Indies and extends as far south as Australia but the trees have been exploited so freely in the past that supplies are now difficult to obtain. It is a slow-drying timber but develops no degrade. Sandalwood is naturally very durable but is used more for making cabinets and purposes where the good heartwood fragrance is an asset than it is for the work where its durability is likely to be of importance; it is also important as an incense wood and for the extraction of Sandalwood Oil used medicinally and in perfumes and cosmetics. It is easy to work and carves and turns well. Sandalwood when dry weighs about 960 kg/m^3. The grain is either straight or wavy and the texture very fine and even; the heartwood of the species is yellowish-brown, deepening after long exposure to the atmosphere to a darker reddish-brown; the wood has a strong distinctive and persistent smell, which is unlike that of any other timber. It is used almost exclusively for small carved items, trinket boxes and other small fancy goods.

Sandalwood, Australian   (Queensland Sandalwood)
Santalum species

Three different species of *Santalum* go to make up Australian Sandalwood, these being *Santalum lanceolatum*, *Santalum cygnonum* and *Santalum preissianum*. Their natural habitat is Queensland, Western and Southern Australia respectively but they may all be marketed together with *Santalum album* (the best known sandalwood), under the general name of Sandalwood. The wood has an average dry weight of about 850 kg/m³, a straight or a wavy grain and a very fine and even texture, the heartwood being yellowish in colour and the timber having the distinctive Sandalwood smell. A considerable export trade is done in this timber from Australia to China, where the wood is prized for joss-sticks and for funeral 'furniture' buried or burned in or near relatives' graves. The tree also yields an oil with a certain commercial importance. Its uses as a timber are confined to turnery, inlaying, fancy articles and so on but the wood is not readily obtainable, even on the Australian market.

Sandalwood, Red   (Red Sanders or Caliatur Wood)
Pterocarpus santalinus

Red Sandalwood has an average weight of about 1000 kg/m^3 when fully dried, a shallowly interlocked grain and a medium fine and even texture. In the heartwood the timber is very dark to almost black in colour, with a figuring of darker streaks in the lighter varieties; quarter-sawn material may have a narrow stripe figuring caused by the interlocked grain. Red Sandalwood is a naturally durable timber. It is not easy to work, whether with hand or with machine tools but it can be brought to a good clean surface and it carves and polishes satisfactorily. In addition to its use as a timber the tree also has medicinal properties and yields a useful dye. The wood is of the same genus as that producing the padauks but it is neither as well known nor as popular as these timbers, nor is it likely to become so. It is very difficult to distinguish from them.

Santa Maria
Calophyllum brasiliense var. *rekoi*

There is also a *Calophyllum brasiliense*, known as Jacareuba, which grows in Brazil. Santa Maria is a timber of the West Indies and Central America and is a tree of good growth habit; large size timbers are obtainable. The two timbers are not generally distinguished from one another.

The sapwood and the heartwood of the timbers are not always well defined; the sapwood is narrow and a pale pinkish shade, while the heartwood is variable from pale pink to rich reddish-brown. The grain is interlocked, the texture moderately coarse and even, while the planed surfaces are mildly lustrous. Dried stock averages out at 610 kg/m³ in weight.

Both timbers shrink considerably in drying and need to be dried slowly to avoid warping; initial temperatures should be kept low in kiln drying to overcome a tendency to splitting round knots. Once properly dried the wood is moderately stable. These are difficult species to plane on the quarter and may need some care in surfacing but otherwise work well, are moderately good turnery species and respond satisfactorily to the normal types of finishing agent. The heartwood is classified as very difficult to treat satisfactorily with preservative fluids but is naturally durable.

The wood is regarded as general utility material – its behaviour in drying spoils it for quality purposes such as furniture. It is used for construction work, flooring and shipbuilding.

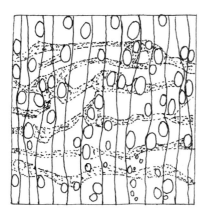

Fairly obvious gum veins are not unusual in this timber.

Sapele
Entandrophragma cylindricum

There are several important species of *Entandrophragma* found in West and East Africa and they are all similar in appearance, although not necessarily in properties. They are closely related to African Mahogany and are sometimes wrongly referred to as mahoganies. Sapele is one of the best known of the species. The heartwood is a medium to dark reddish-brown, quite distinct from the pale creamy yellow sapwood. The wood nearly always shows a pronounced inter-locked grain which yields a beautifully regular stripe figure when quarter sawn or sliced. It is also seen occasionally with wavy grain which will lead to a mottle or fiddleback figure. The texture is modera-tely fine and the weight about 620 kg/m³ dried. Freshly cut stock has a short-lived cedar-like odour. It is somewhat difficult to season without distortion.

Movement   –   medium
Durability   –   moderately durable
Treatability   –   resistant including the sapwood
Workability   –   works well except for planing quartered stock; finishes well.

Sapele is as strong as Oak in most respects and stains and polishes so well that it is much used for furniture both as solids and as decorative veneers. It is also peeled for plywoods and used in cabinet work, interior fittings, boatbuilding and flooring.

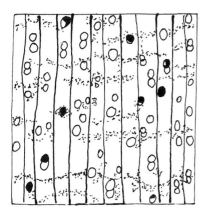

Ripple marks are invariably present. The vessels are smaller and more numerous than in Utile (with which it is often confused).

(continued overleaf)

237

Another species, *Entandrophragma candollei* is known as Omu or Heavy Sapele – but it is only heavier when green. The wood is generally coarser, darker and less attractive than Sapele, although it may be used for similar purposes. *Entandrophragma angolense* is known as Gedu Nohor or Edinam. Again it is similar to Sapele but is coarser and not well interlocked, so produces much plainer and often paler timber. It is also quite a lot lighter in weight, averaging 540 kg/m³ dried. It can be used as a plainer alternative to Sapele or instead of African Mahogany. Utile is the timber of *Entandrophragma utile* which has a separate entry.

Sassafras
Sassafras variifolium

In many respects this tree is valued more for its by-products than its timber, and from it come bark and root beer, patent medicines and an essential oil that is the source of artificial heliotrope used in perfumery and the manufacture of cosmetics. The timber is refractory under drying treatments and is subject both to insect and fungal attack. In all hand and machine processes the wood works well, with no particular dulling of tools. It has a lightish red-brown coloured heartwood, a moderately coarse and uneven texture and a grain that is generally straight. The wood is quite light in weight, at about 500 kg/m^3 when dried, has a characteristic strong but pleasant smell and is one of the relatively few timbers that show a ring-porous structure. A considerable quantity of the wood reaches the market in the form of small chips for use in the dyeing industry. Two other hardwood species known as Sassafras are the Yellow Sassafras (also known locally as Golden Deal), the timber of *Dryophora sassafras*, native to Queensland and New South Wales and the Tasmanian Sassafras (*Atherosperma moschatum*). Both timbers have, in general, good properties but are of purely local importance.

Satinwood, Ceylon   (East Indian Satinwood)
Chloroxylon swietenia

The finest timber of this species comes from Sri Lanka, and the wood is well known but scarce on the world markets. It has an average weight of about 960 kg/m^3 dried and a fine and even texture. The heartwood is light or golden yellow in colour, quarter sawn stock showing a narrow stripe figuring often combined with some wavy grain which results in an attractive mottle figure and it has a very lustrous surface. The timber is naturally durable when used in exposed positions but is regarded as a cabinet wood, not being used externally to any great extent; it is difficult to impregnate with wood preservatives. Drying defects are quite common unless it is dried in log form and it is not easy to work, though it can be brought to a good clean surface; its turning and carving qualities are good and it polishes well. Amongst the typical uses to which the timber is put are cabinet making, turnery, small fancy articles, carving, panelling and furniture. It is used in veneer form and is often seen in antique furniture. Ripple marks are present but difficult to see. The structure is very fine generally and little can be seen apart from narrow lines of parenchyma marking the growth rings.

Satinwood, West Indian
Fagara flava

West Indian Satinwood is quite well known on world markets and is the product of trees from the West Indies. It is not durable and finds its greatest use for internal work. It works quite well, under either hand or machine tools and also turns satisfactorily but for a cabinet timber it suffers from the disadvantage that it has a high oil content and polishing processes do not always produce satisfactory results. The wood weighs about 880 kg/m^3 dried. The grain is irregular or interlocked and the texture fine and even; the heartwood of the species is yellowish in colour, darkening to golden yellow after prolonged exposure to the atmosphere. West Indian Satinwood has a characteristic taste and a distinctive smell of coconut oil when fresh. It may cause dermatitis among operatives machining the timber if careful attention to dust extraction is ignored. Ripple marks are not present in this species.

Sepetir (Makata)

Sindora species

There are several South East Asian species of *Sindora* which are not separated. A very similar timber known as Swamp Sepetir (*Pseudo-sindora palustris*) also exists and can easily be confused with *Sindora* species. Swamp Sepetir comes only from Sarawak and Sabah.

Sepetir is a medium brown wood, sometimes with dark streaks – sapwood is distinct and may be very wide. The grain is variable, often shallowly interlocked and quartered surfaces may be quite decorative. The texture is fine and even – terminal parenchyma is prominent and gives rise to a growth ring figure on surfaces. The dried weight about 670 kg/m³. Fresh timber smells spicy.

Movement	–	small
Durability	–	durable
Treatability	–	very resistant
Workability	–	hard to work but finishes well; needs pre-boring; may be a little sticky from resin on first cutting.

The wood is used for joinery and light construction. Figured logs produce good decorative veneer for furniture.

Swamp Sepetir is much plainer and rather more red than Sepetir but is otherwise similar.

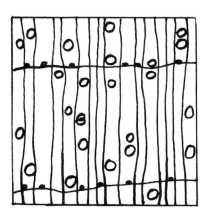

The terminal parenchyma is obvious but the vertical resin canals associated with it are not easy to see. Swamp Sepetir has no canals.

Seraya, White (Bagtikan)
Parashorea species

Detail of the various serayas and merantis was given under the entry for Lauan. White Seraya does not fit the scheme properly in that it is not the equivalent of White Meranti but is the product of only two species of *Parashorea* – it is therefore more constant in its properties. If anything it is more the equivalent of White Lauan in that it weighs about 530 kg/m^3 when dried. It grows in Sabah and the Philippines.

The wood is mostly rather straw coloured, sometimes very pale brown or pink tinted. The texture is rather coarse and the grain usually interlocked.

Movement   –   small
Durability   –   non-durable
Treatability –   very resistant
Workability   –   not difficult but needs sharp tools to avoid a fluffy finish

This timber is used for interior work, domestic flooring and plywood. It has been used for ship decking because of its light colour but is not recommended because of its lack of durability.

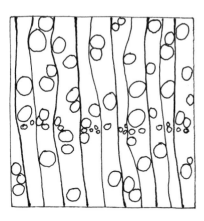

Care is needed to avoid brittleheart in this species. The lines of ducts are usually easy to see.

Silky-oak, Australian   (Northern Silky-oak)
Cardwellia sublimis

The name Silky-oak is used because the wood has large rays made of lustrous parenchyma tissue which make it look a little like Oak. It is not a true oak, however, and the name Silky-oak may also be applied to a related species, *Grevillea robusta*. Both species grow in Australia and New Zealand – *Grevillea robusta* is also imported from East Africa where it is grown as a coffee and tea plantation shade tree; in order to distinguish this species it is referred to as Southern Silky-oak in Australia and Grevillea in the U.K.

Cardwellia sublimis is quite a reddish-brown when first cut but becomes a darker brown on exposure. It is a coarse-textured wood, usually straight-grained and weighs about 530 kg/m³ when dried. Silver figure is very obvious on quartered surfaces.

Movement   –   probably medium
Durability   –   moderately durable
Treatability   –   probably moderately resistant
Workability   –   easy to work but not easy to get a good finish owing to the large soft rays.

The wood is used in the solid and as a veneer for furniture, panelling and shop fitting. Grevillea is paler and generally rather inferior in quality although basically similar.

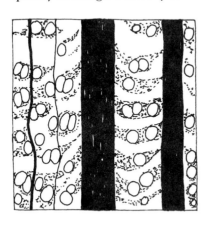

The looped arrangement of vessels is characteristic. Rays of two sizes are present. White deposits are common in *Cardwellia* but absent in *Grevillea*.

Silver-grey Wood, Indian   (White Chuglam)
Terminalia bialata

This name is given to the figured timber of *Terminalia bialata*; the plain timber is described as White Chuglam: the properties of the two varieties are so similar that one description will cover both. The timber comes from the Andaman Islands and Burma and can be obtained in very good sizes. The sapwood of the timber is of a creamy yellow shade when fresh, though it eventually darkens slightly. In Chuglam the indistinguishable heartwood colour is a greyish-yellow; in Indian Silver-grey Wood there is additional marbling of irregular darker markings which is attractive. The wood is lustrous, has a straight grain, medium texture and weighs about 670 kg/m^3 when dried.

Movement   –   no information
Durability   –   probably moderately durable
Treatability   –   very resistant
Workability   –   fairly easy to work and finishes well

This decorative wood is used for furniture, panelling, cabinet making and good quality joinery. It is used both in the solid and as veneers. White Chuglam is not usually exported from India.

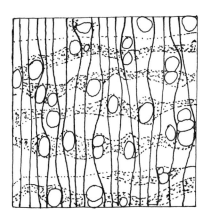

This species of *Terminalia* shows tangential bands of parenchyma very strongly developed. In other species these bands are absent or very thin.

Snakewood (Letterwood)
Piratinera guianensis

Snakewood is also called Leopard Wood or Tortoiseshell Wood. It is very heavy, having an average weight in the region of 1300 kg/m³ when dry; it has a straight grain, a fine and even texture and a heartwood that is dark red or reddish-brown in colour, with darker markings resembling a snake skin, or a leopard's spots. Only the heartwood of this Tropical American species is of any commercial use but the timber provides one of the heaviest of all timbers. The sapwood may be very large and is normally cut off – only the heartwood is exported. Unlike most timbers Snakewood is sold by weight. The timber is difficult to work with hand tools and is somewhat brittle, needing pre-boring for nailing, mortising, etc. The timber slices well for veneers and finishes well, polishing exceptionally well. Another timber known as Snakewood comes from *Strychnos nuxvomica*, also known as the Strychnine Tree or Nux-Vomica. The tree yields strychnine and is of considerably more medicinal than mechanical importance, though the wood has a bitter taste that renders it immune to termite attack. Average dry specimens weigh from 800 to 900 kg/m³ and have a heartwood that shows a distinct yellowish tinge and possesses a fine and even texture. As a timber it is only of very rare use and its economic importance is negligible.

Sneezewood
Ptaeroxylon obliquum

This hard and strong timber is a member of the same botanical family as that producing the true mahoganies but the wood is used for very different purposes. Its principal uses are for fencing posts, piling, poles and other external work where its great resistance to decay will stand it in good stead. It is a wood with a distinctive smell that is strongly reminiscent of pepper. Average timber is irregularly grained and coarse and unevenly textured, weighs between 900 and 1000 kg/m^3 dried and when freshly felled has a bright red heartwood that later tones down to a brownish tint. It is not difficult to dry whether in air or by kiln treatment but is rather awkward to machine as the sawdust has irritant properties and causes considerable sneezing whilst the wood is being worked. Sneezewood needs pre-boring before nailing, screwing, or similar operations and, having a high oil content, does not always glue well.

Sterculia

Sterculia species

There are two commercial species of *Sterculia* available from West Africa: *Sterculia oblonga* has a yellowish to pale yellow-brown heartwood and is known as Yellow Sterculia; *Sterculia rhinopetala* has a pale to deep reddish-brown heartwood and is known as Brown Sterculia. In the case of Yellow Sterculia the sapwood is wide and not easily recognizable but with Brown Sterculia the sapwood is of a lighter colour that marks it off clearly from the heart. With both varieties the grain is usually interlocked (occasionally straight in Brown Sterculia) with a texture that is rather coarse. Quarter-sawn stock shows as well marked silver figure. Brown Sterculia averages out at 820 kg/m³ and Yellow Sterculia at 780 kg/m³ when dried. Sterculia dries only slowly and has some tendency to warp and split: movement may be quite severe when it is dry. The wood is not very resistant to fungal attack and the heartwood is exceptionally difficult to treat satisfactorily with preservative fluids. The wood has a tendency to pick up in surfacing but otherwise it is not difficult to work. It may need considerable filling before polishing. Both varieties are used chiefly for structural work, flooring, roofing shingles, etc., though occasionally Brown Sterculia may be used for furniture and interior joinery.

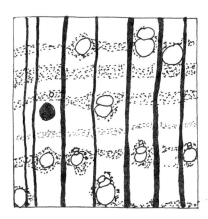

Some gum and other deposits may be present in the vessels of Yellow Sterculia but are absent in Brown Sterculia. Heartwood colour is the most reliable distinguishing feature.

Stinkwood

Ocotea bullata

(Cape Laurel)

Stinkwood is a timber from South Africa. The heartwood ranges from pale yellow through shades of brown to almost black in colour, sometimes showing a striped figure on quarter-sawn surfaces. The grain is interlocked and the texture reasonably fine and even; the average weight is about 800 kg/m³ when dried; the characteristic unpleasant smell that has given the wood its name is not discernible in the fully dried timber and is therefore no great drawback. It is a tough and strong wood that has long been well known for its beautiful and distinctive figuring but supplies are no longer readily available and with the demand greater than the possible supply, the timber commands a high price. Stinkwood is a difficult timber to dry, and degrade during that process may be severe. It is not easy to work but good results can be obtained if the cutting edges of tools are kept sharp. The wood is used only for such purposes as high-grade furniture, panelling, ornamental turnery and similar items.

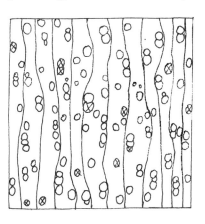

Tyloses are usually present in the heartwood; otherwise the structure is unremarkable.

Sweetbark
Pradosia latescens

This is a tough and elastic timber coming from a tree native to Brazil and the neighbouring regions. It is of local importance only but on account of its resilience it is said to be used as a substitute for Ash or Elm. The tree is normally of tall growth habit, so good-sized timbers are not uncommon; the timber has a sweet distinctive taste that has been responsible for its name. The wood is generally straight in the grain and has a medium or fine texture, whilst its weight is about 960 kg/m³ when dried. The sapwood and the heartwood of average samples of the timber are not at all well delimited, the timber being a whitish or a dull grey in tone, with no distinctive figuring. It is said to be a reasonably durable wood for use in positions where it may be attacked by a wood-rotting fungus but it suffers far more readily from sap-stain fungal attack. Reports seem to indicate that the wood works easily and finishes well.

Sycamore

(Sycamore Plane, Plane)

Acer pseudoplatanus

This tree is the same genus as that providing the maples and the timbers are often confused. Sycamore is native to Europe and Western Asia. Further confusion can be caused through use of the alternative names Sycamore Plane and Great Maple in Britain and Plane in Scotland.

Sycamore is a pale slightly yellowish-white lustrous timber with no distinction between heartwood and sapwood. The grain is mostly straight but may be wavy and so gives rise to the attractive fiddleback figure traditionally used for violin backs. The texture is fine and even and the weight about 610 kg/m^3 when dried. The wood stains quite badly unless it is dried quickly.

Movement	–	medium
Durability	–	perishable
Treatability	–	permeable
Workability	–	works easily and finishes well – wavy grain needs special care.

The wood is strong and turns very well. It is used for all kinds of bobbins, handles and rollers; for domestic and dairy utensils; for veneers when figure is present. The wood can be chemically treated to make it a soft grey colour – it is then sold as Harewood which has been used for panelling and marquetry.

Tallowwood
Eucalyptus microcorys

The tree producing this timber is native to New South Wales, Fraser Island and Queensland, the timber being well known and popular on the world markets; it is related to Jarrah. The wood is remarkably durable when used in exposed positions and is classified as impermeable to treatment with preservative fluids. It is not difficult to season, though fine cracking may develop during the process and once dried it has a high degree of stability. Tallowwood is strong and somewhat difficult to work, either in hand or machine processes and will not glue well on account of its greasy nature. The average weight is about 990 kg/m³ when dried; the grain is normally interlocked and the texture moderately coarse but even. The sapwood and the heartwood of the species are not well defined, the timber being dull yellow to light brown in colour, sometimes showing a darker figuring but it has a uniform tint in any given specimen; the machined wood has a lustrous surface and a greasy feel which accounts for the common name. Tallowwood is used for harbour and dock work, flooring and some structural work.

Tawa
Beilschmiedia tawa

Tawa is an important New Zealand timber. It is a pale yellowish white with no distinct heartwood and is very similar in appearance to Sycamore. The texture is fine and even, the grain usually straight; the timber weighs about 750 kg/m³ when dried. It works quite well in most operations, though some care may be needed in planing to counteract a tendency to picking up; pre-boring for nails or screws is advisable with the thinner dimensions or the timber may split. The drying and strength properties are good and Tawa will hold its shape well when dry. Although sufficiently durable for all normal uses under cover, Tawa may be attacked by furniture beetle and is not resistant to fungal attack; it is unlikely to be given a preservative treatment in view of the uses to which it will be put. Tawa is a good flooring timber with high resistance to wear. It is also used for furniture and joinery in New Zealand.

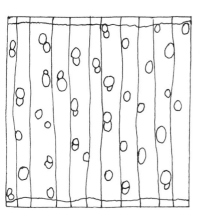

The structure is similar to that of Queensland Walnut with a faint parenchyma around the vessels and forming terminal lines.

Teak is a well-known timber imported from the South East Asian region and may be described as Moulmein Teak, Rangoon Teak and so on, according to its port of shipment or neighbourhood of origin. The tree is grown in plantations in East and West Africa and the West Indies. The name African Teak may be used for real Teak, or for Iroko or sometimes Afrormosia, so care needs to be taken. Teak weighs between 610 and 690 kg/m³ when dried. The heartwood is a golden brown colour which may darken considerably on exposure. In some material darker brown streaks are present and they remain as a kind of permanent figuring. The wood has an oily feel and unless recently surfaced this may give a problem with adhesives. There is also a characteristic smell in fresh material, said to resemble that of leather – this disappears on drying. The grain is mostly straight and because the wood is more or less ring-porous there is usually a fairly strong growth ring figure.

Movement	–	small
Durability	–	very durable
Treatability	–	very resistant
Workability	–	generally not too difficult except with the heavier material; needs pre-boring; very blunting to cutters.

This wood has a combination of good properties that have made it popular. Typical uses for Teak include the making of furniture, acid tanks, boxes and chests for tropical use and it is also favoured for use in ship-building for decking and similar work. It is very good for external joinery and garden furniture in spite of its cost. It is attractive as flooring but not hard wearing – it is best when quartered.

Philippine Teak is a true teak inasmuch as it comes from *Tectona philippinensis* but it is of very limited occurrence and little commercial importance. In colour it is rather lighter than Burma Teak and has a grain that is usually slightly wavy but in its other gross features and qualities it closely resembles its better-known relative. In the Philippines the wood is used, when available, as a structural wood rather than as a cabinet timber. The name of Teak is often given to timbers not of the true Teak family (the Verbenaceae) and often of local importance only. Care should be taken to distinguish between such timbers and the true Teak.

Teak *continued*

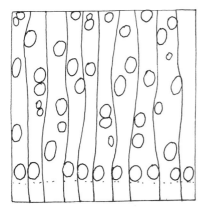

The ring-porous structure either is well developed or consists only of a few large vessels in a broken line.

Teak, Malacca
Afzelia species

This timber comes from trees native to Borneo, Malaysia, Thailand, Sumatra and other neighbouring regions of Asia and is a hard and durable wood that shows considerable resistance to termite attack, though it does not stand up so well to marine borers. It is rather difficult to work and tools need constant oiling as they are apt to clog because of the thick gum deposits in the wood; it can be surfaced satisfactorily and will polish well. The timber contains an agent that, like common Oak, causes the corrosion of ironwork fastened to it. Malacca Teak finds use for such items as flooring, railway sleepers and as a structural wood and is also of some importance in the dyeing trade. It is somewhat heavier than true Teak, averaging 900 kg/m^3 when dry, is normally straight-grained and has a coarse texture and a dark yellowish-brown heartwood. The timber is unlikely to achieve any important position on the general timber markets, although it will always be fairly important locally.

Teak, Rhodesian   (Zambesi Redwood)
Baikiaea plurijuga

Rhodesian Teak comes from a tree which grows in Southern Africa. The heartwood is reddish-brown in colour with irregular darker markings, the sapwood pale and clearly distinguishable. The heartwood is often pale on surfacing but darkens in a few hours. The grain may be either straight or shallowly interlocked, the texture fine and even; the weight is about 960 kg/m³ when dried. The tree is not of tall growth habit (hence timbers are not of large dimensions) and reaches its best development on river banks. The wood is not difficult to dry, surface checking being almost negligible, though a certain amount of warping may occur. Evidence of wood-borer attack may be found in converted wood but the timber is naturally very durable against decay and shows a marked resistance to termite attack, being similar to the true Teak in this respect. Rhodesian Teak is difficult to work and dulling of tool edges may be quite appreciable, though it may be finished to a smooth surface, will turn extremely well and polishes very satisfactorily. The wood is well known, being regarded as a good substitute for and used for much the same purposes as true Teak. It is particularly used for flooring.

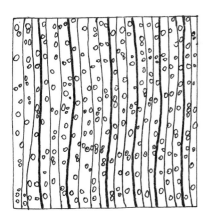

The vasicentric parenchyma is usually a little hard to see. Ripple marks are present.

257

Thingadu

(Kaunghmu)

Parashorea stellata

The timber is the product of a tall tree of Burma and the Malay Peninsula; the wood is not well known on the world markets, though it can be obtained in fair-sized dimensions. It is mildly decorative in appearance and is therefore suitable for panelling, cabinet making and similar work. Under drying treatments it is rather apt to stain and warp. It works well however, can be brought to a good surface and finishes well after filling. It is only moderately resistant to the attack of decay. Reddish-brown in colour, with quarter-sawn stock showing a ray figuring, the timber has an interlocked grain and a very coarse but even texture, its weight being 710 kg/m^3 dried. No demand is likely to arise for the wood on the world market but a local increase in demand is not unlikely. Axial gum canals are present. The wood is suitable for boat building, flooring and general construction.

Thitka

(Kashit)

Pentace burmanica

This timber, sometimes known as Burma Mahogany, has an average weight of about 690 kg/m³ when dried. The grain of the wood is normally interlocked, the texture rather coarse and even; the heartwood is red-brown in colour, darkening with age and exposure to the atmosphere, with a stripe figuring to be seen on radially sawn surfaces. The tree producing the timber is common in Burma, the Malay Peninsula and Java, though only in the first named of these countries is it commercially exploited. The timber behaves fairly well under drying treatments and degrade is not excessive. It is a naturally durable timber when exposed to the weather or if in contact with earth but is markedly less resistant to termite attack. The only difficulty likely to be encountered in the working of the timber is in surfacing quartered material. It finishes well. Typical uses that may be quoted for Thitka include furniture manufacture, interior fittings, flooring, boxes, piano cases, mathematical instruments, boat-building, shop fittings and similar items. Ripple marks are present and usually very easily seen.

Tulipwood
Harpullia pendula

Tulipwood, or Tulip Lancewood as it is sometimes called, is a tree of limited availability and it is not easily exploited. The timber itself is a heavy one, averaging about 960 kg/m^3 when dried and it has a grain that may be either straight or shallowly interlocked, whilst the texture is normally fine and even. The heartwood and sapwood of the species are sharply differentiated, the latter being broad and whitish in colour, whilst the heartwood has a dark brown background with a tortoiseshell figuring of darker markings. There is no distinctive taste or odour to the timber. The wood is only available in small dimensions but is obtainable on the market in small consignments, though the price is high. Tulipwood is said to be naturally durable for use in conditions favouring fungal attack but is regarded as a cabinet wood rather than as a structural timber, being used for the making of small fancy articles and novelties, mouldings, walking sticks, tobacco pipes and similar items. A steady small demand exists for the wood. This Australian wood should not be confused with the American Whitewood, sometimes called Tulip Tree or Tulipwood.

Tupelo
Nyssa species

Nyssa aquatica may be sold as Water Tupelo or Tupelo Gum; *Nyssa ogeche* as Tupelo and *Nyssa sylvatica* as Black Gum or Black Tupelo. The name Cotton Gum may also be used. The species all grow in Southern and Eastern states of North America.

The species are all similar. The sapwood of Tupelo is wider than most timber species. It is whitish or grey in colour and merges only gradually into the light brown heartwood which is sometimes streaked. The grain of the wood is interlocked (giving quarter-sawn stock a ribbon stripe figure) but otherwise it is rather featureless. The texture is close and even, the wood has no appreciable smell and when dried weighs about 510 kg/m^3 – *Nyssa sylvatica* is a little heavier at 560 kg/m^3.

Tupelo tends to warp unless care is taken during drying and the interlocked grain may be difficult to finish. The wood is rather difficult to work, in general being soft and light. It is, however, quite tough and resistant to wear so it is used for flooring in heavy use situations, as well as general construction and interior use. Preservative treatment is needed under exposed conditions. It is also used for packaging and core stock in certain plywoods.

Turpentine (Luster)
Syncarpia glomulifera

This Australian timber is the product of a species from Queensland and New South Wales. The wood is somewhat refractory under drying treatments and collapse is a common condition but the timber may be restored to its original size, shape and quality by steam reconditioning treatments. Turpentine is hard and dense and is naturally resistant to termites and marine borers as well as to the onset of decay; it is also said to be resistant to fire hazards. The heartwood of the species is reported to be impermeable to treatment with preservative liquids. The wood has a gritty surface but is not really difficult to work with machine tools and it may be turned and bent satisfactorily; it will polish well. Average specimens of the timber weigh about 950 kg/m³ when dried, the grain is interlocked and the texture medium coarse but even; the heartwood of the species is reddish-brown in colour, whilst quarter sawn stock will show a slight ray figuring. Among the uses to which the wood may be put railway sleepers, dock works, wagon- and ship-building may be mentioned.

A related tree is *Syncarpia hillii* which produces a timber called Satinay. It has an average weight of about 900 kg/m³ when dried, an interlocked grain, a fine and even texture and a heartwood that is reddish to reddish-brown in colour; quarter-sawn stock may have a striped figuring. It is a moderately hard, moderately durable timber that is the product of a tree native to the Australian continent that reaches its best development on Fraser Island, off the coast of Queensland. The timber is said to have received its name because of its resemblance to the so-called 'satine' timbers of tropical America. The wood is of importance to Australia only and although sufficient quantities of the timber probably exist to meet all requirements that may arise, a greatly increased demand is very unlikely. Satinay may be used for panelling, cabinet making, veneering and structural purposes.

Utile (Sipo)

Entandrophragma utile

This species, another member of the mahogany family, is very like Sapele and is often confused with it. Utile grows in West and Central Africa and produces a medium reddish-brown timber of moderately coarse texture and irregularly interlocked grain. Quartered surfaces show an irregular and far less attractive stripe figure. The dried weight is about 660 kg/m³. Distortion in drying is not severe.

Movement   – medium
Durability   – durable
Treatability   – very resistant
Workability   – works well except for planing quartered stock; finishes well

Utile is regarded as a better-behaved timber than Sapele in that it dries more easily and is less likely to move in use. It can be used as solids together with decorative veneers of Sapele to secure the best of both species. Utile is used for furniture, joinery, cabinet work and boat-building.

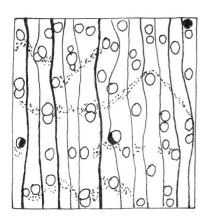

The parenchyma bands are usually wavy in Utile and much straighter in Sapele. Ripple mark is often absent.

Vinhatico   (Vinhatico Castanho)
Plathymenia reticulata

When freshly sawn this timber has a distinctive yellow or orange-brown heartwood, sometimes with darker streaks, though the timber will eventually fade to a rich dark shade of reddish-brown; the sapwood is only narrow and is clearly defined. The grain is variable from straight to roey, planed surfaces are rather silky, the texture is moderately fine and even, while the average weight is about 600 kg/m^3 when dried. Vinhatico is a Brazilian timber. The timber is not particularly well known in Britain but may be used for furniture, cabinet work, shoe heels and joinery. It is easy to work and to dry. When dry it is stable but difficult to polish unless a grain filler is used.

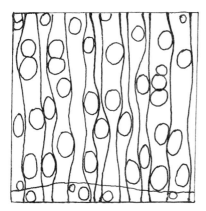

Ripple marks are often present but not in all material.

Virola (various)
Virola species

The species of *Virola* growing in Central and South America produce timbers of varying colours and weights.

It is preferable to distinguish the medium to dark brown timbers with an average dried weight of about 670 kg/m³ as Heavy Virola. Typical of this type is *Virola bicuhyba*, also known as Becuva or Bicuiba.

Pale brown or pinkish brown woods with an average weight of 430–580 kg/m³ when dry are referred to as Light Virola. Typical of this type are *Virola surinamensis* and *Virola koschnyi*, also known as Banak, Baboen or Ucuuba.

All the species have straight grain, medium coarse and even texture and need some care to dry them thoroughly.

Heavy Virola is moderately durable, stable, works easily and finishes well and is used for furniture, carpentry and construction.

Light Virola is perishable, inclined to stain if not dried rapidly, works very well and finishes well. It, too, is stable when thoroughly dry and is used for plywood and a wide variety of light constructional purposes. It can be easily stained to resemble mahoganies.

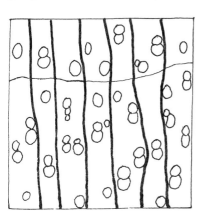

Light Virola sometimes has terminal parenchyma, otherwise the two types are distinguished on the basis of weight and colour. Scalariform perforations are present.

African Walnut may also be known as Nigerian Walnut or Benin Walnut. It has an average weight of about 560 kg/m³ when dry. The grain is normally interlocked, with the texture fine and even, while the heartwood is a golden- or yellowish-brown in colour, often having a figuring of darker markings which are actually gum canals, with a striped figure on quarter-sawn surfaces. The timber dries well with only a low rate of degrade.

The wood is not a true walnut and does not show the slight greyishness that walnuts have. It is in fact a member of the mahogany family and is not at all unlike them.

Movement   –   small
Durability   –   moderately durable
Treatability   –   very resistant
Workability   –   works easily except in planing quartered material but needs sharp tools to obtain a good finish

The wood is a good timber in its own right and does not need to pretend to be a walnut. It is a handsome furniture timber, both as solid and veneer, and is also used for doors, shop fitting, high quality interior joinery and cabinet making.

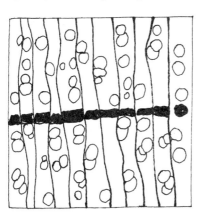

The gum ducts or canals are caused by insect attack to the standing tree and are not natural ones as found in Meranti. They are called 'traumatic' canals.

266

Walnut, American   (Black Walnut)
Juglans nigra

The tree producing this timber reaches its best development in the Ohio River Basin and under favourable conditions may reach a height of 30 metres; it is a member of the true walnut genus. The timber is well known and popular on the world markets but it has been so freely exploited in the past that supplies tend to diminish year by year and consequently become rather expensive. It is more durable than English Walnut, though it is used for much the same purposes as that timber, that is for high quality cabinet making, veneers, furniture, internal fittings, mouldings, plywood, piano cases and similar items. The average weight of the wood is about 640 kg/m^3 when dry, the grain is generally straight and the texture coarse; it is stable after drying. The heartwood of the species may vary from a chocolate brown to a deep purple in colour – in general the colouring is more uniform than in European Walnut. Like European Walnut it is also used as a standard timber for rifle and gun stocks.

Another species *Juglans cinerea*, also grows in North America; it produces a paler, weaker, softer timber which is known as Butternut or White Walnut. This is generally inferior to other true walnuts and is used for lower quality goods and less exacting purposes. It weighs about 430 kg/m^3 dried.

Structurally the walnuts are all very similar; separation on the basis of colour and weight is reasonably reliable.

Walnut, Blush
Beilschmiedia obtusifolia

Blush Walnut is a rather heavy timber (average weight about 875 kg/m³ when dried), has a straight grain and a texture that is variable from fine to medium coarse but even; the heartwood is yellowish-brown to pinkish-brown in colour. The timber comes from an Australian tree found in Queensland and New South Wales but it is not of the true walnut family. It is a moderately durable wood that is best suited for use in protected positions and is accordingly used for such items as flooring and interior fittings. The timber turns well and might be found outside Australia in this form. The wood may be regarded as being primarily of local interest only, with little likelihood of any demand arising for it apart from local consumers.

Juglans regia

Walnut is one of the best known hardwoods in the world and is deservedly popular as a high-class furniture wood. The tree is common in southern Europe and parts of Asia, being extensively cultivated for its fruits; the valuable timber also reaches the market under such descriptions as French, Spanish, Circassian, Italian, Turkish or Persian Walnut.

The wood is quite variable in colour. The sapwood is always very light and easily distinguished; the heartwood is usually greyish-brown with irregular darker streaking. The amount of streaking is also variable in its distribution and its intensity; streaks may be confined to the centre of the log only. The place of origin will affect the appearance, so it is possible to say that French Walnut is plain and grey while Italian Walnut is darker and strongly streaked with black and brown – such strongly figured wood is often referred to as Ancona Walnut. After long exposure the wood assumes a golden brown colour in which the grey is often not obvious. This tree often has burrs or areas of wavy grain and these are available as veneers which can be matched and patterned to provide particularly beautiful effects for furniture and cabinet making. The grain is mostly straight, the texture moderately coarse and the weight is about 640 kg/m³ when dried.

Movement	–	small
Durability	–	moderately durable
Treatability	–	resistant
Workability	–	easy to work and finishes well; moderate blunting effect on tools; very good for steam bending

European Walnut is used both in the solid and as veneer for high quality furniture and cabinet making. It is also popular for interior fittings and for a variety of turned goods. It is the preferred timber for rifle and gun stocks because of its stability. The wood can stain when in contact with iron under damp conditions.

(continued overleaf)

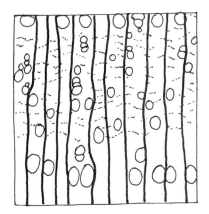

The species usually shows a semi-ring porous structure with only a few large vessels in the earlywood of each growth ring. The fine lines of parenchyma are not always easy to see.

In addition to the American Walnuts described earlier, the Japanese Walnut (*Juglans sieboldiana*) also has a certain amount of importance as a timber, although rather pale in colour.

Among the timbers also described as walnut (though not members of the walnut genus) may be mentioned Brazilian Walnut (or Imbuya) which comes from the genus *Phoebe* and is closely related to Greenheart. Brazilian Walnut finds use locally as a sleeper wood and structural timber, as well as a cabinet wood. Many timbers with a greyish or striped appearance have been referred to as walnuts at various times.

Walnut, Queensland

(Australian Walnut)

Endiandra palmerstonii

This North Queensland timber (also known on certain markets as Australian Black Walnut or Walnut Bean) is one of the most popular Australian hardwood timbers to reach the general consumer and it is quite freely exported. It is not a true walnut. It has a grain that is either interlocked or wavy, with a medium and even texture. The weight on average is 680 kg/m^3 when dried. The wood is pale to darkish brown with darker irregular streaks, with quarter sawn timber showing a broken striped figuring; the wood has a mild unpleasant smell before drying. Queensland Walnut is difficult to work but peels or slices satisfactorily for veneers and is capable of taking a high polish. It has a severe blunting effect on cutting edges. The timber is rather difficult to handle under drying treatments and is not naturally durable when used in situations unprotected from the weather. It is a typical cabinet wood, being used for such purposes as turnery, cabinet making, furniture, interior fittings, panelling, veneering and flooring.

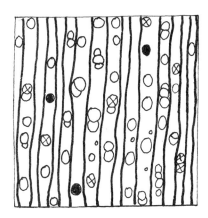

Dark brown deposits and tyloses are common.

Whitewood, American   (Tulip Tree, Canary Wood)
Liriodendron tulipifera

This is best described as a rather featureless general-purpose utility wood. It is also sometimes called Basswood but this confuses it with *Tilia americana*.

This once well known timber is the product of a tree native to the eastern half of the North American continent and was readily obtainable on the world markets in the form of wide boards. The wood is rather soft and works well under hand or machine tools, being especially useful for carving. Whitewood will not polish satisfactorily but takes paint and stain well. Light in weight, the wood averages from 450–510 kg/m^3 when dried, the texture is fine and even and the grain usually straight; the heartwood is a pale yellowish-brown, brownish or almost purple in colour, the sapwood being whitish to pale yellow.

The wood is of considerable economic importance in America but is no longer much seen in the United Kingdom. It is suitable for non-show wood in furniture and interior joinery and is also widely used for plywood.

American Whitewood has a very fine structure, not unlike that of Lime, except that very distinct lines of terminal parenchyma are present. Scalariform perforations are also present.

Willow

Salix species

Various species form the Willow of commerce in the United Kingdom. *Salix alba* provides White or Common Willow, *Salix fragilis* provides Crack Willow and *Salix alba* var. *calva* provides the Cricket Bat Willow. All grow in the British Isles, Europe and Asia, mostly in wet areas and close to rivers and streams. Many Willows are pollarded to force them to produce many thin branches to be used in basketry and hurdle fencing.

The wood varies greatly in quality according to the place and style of growth. It is generally a very pale pinkish-white and is very similar to Lime or Poplar. It weighs only about 450 kg/m³ when dried and is soft and not very strong.

Cricket Bat Willow is used to make bats of all qualities and is usually carefully grown. The other species are very variable and are used for artificial limbs, toymaking, trugs, fruit and vegetable baskets and boxes and general crate work.

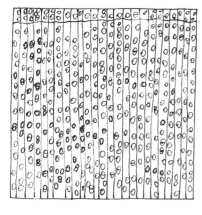

Like Poplar, Lime, Horse Chestnut and some other timbers this timber shows little detail of structure and is not always easy to identify with certainty.

273

Yellowwood
Various species

As is the case with so many other timbers, Yellowwood is a very common name applied to vastly different species. Two varieties of these are dealt with below.

Australian Yellowwood is the timber of *Flindersia oxleyana*, which is a pale yellow-brown in colour, with no characteristic taste or smell, a grain that is generally interlocked and a fine and even texture; average specimens weigh up to 800 kg/m^3 when dried. It is of more importance locally than worldwide. This particular wood works reasonably well in all hand or machine operations and bends satisfactorily but is not naturally durable when exposed to the weather or in contact with the soil. It is used in Australia for such purposes as flooring, linings, tool handles and bentwood work.

The American Yellowwood is the timber of *Cladastria tinctoria* and may be sold in North America as Gopher Wood. It has the yellow tint characteristic of all timbers sharing the vernacular name of Yellowwood; the working and drying qualities are quite good but the wood is of little commercial importance.

Zebrawood
Various species

Locally this is not an uncommon name but only two of the many timbers given this title will be dealt with here. The first of these timbers comes from French Guyana where it is probably better known as Bois Serpent, though it is also given the title of Surinam Snakewood. It is the product of *Pithecolobium racemiflorum*. Occurring only rarely in the forest, the tree is of local importance only, although it provides a useful cabinet wood. It is a very heavy timber even when fully dried, is straight grained and shows a medium coarse texture. In colour the wood is light brown with an irregularly striped figuring. In all tool processes the wood cannot be described as easy to work but it may be brought to a good surface and finishes well.

The second species of Zebrawood is a little-known cabinet wood which is the product of a small tree, *Connarus guianensis*. It is a lightweight wood that has good working and finishing qualities, though it is only moderately durable. It is chiefly noteworthy for its beautiful and distinctive figuring, consisting of a striping of dark reddish-brown on a creamy background.

The name Zebrawood may also be applied to Gonçalo Alves (*Astronium fraxinifolium*) and to Andaman Marblewood (*Diospyros marmorata*) both of which have separate entries.

4 Softwoods

In Chapter 3 many small drawings of end-grain structure were provided as an aid to identification but they are not helpful for softwood timbers. There is very little variation in the structure of softwoods except for the quantity and visibility of the darker latewood in each growth ring. Larches, spruces, pines and Douglas Fir all have resin canals which show on the end-grain quite clearly when using a hand lens.

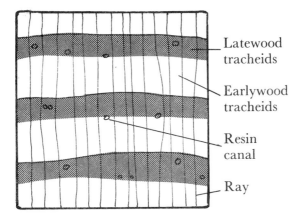

Latewood
tracheids

Earlywood
tracheids

Resin
canal

Ray

This picture shows a typical piece of Scots Pine or Redwood (*Pinus sylvestris*). About one third of the ring is latewood which shows up well as a darker region. The resin canals are mostly near the boundary of early/latewood. In all softwoods the rays are very narrow and therefore no help in identification.

In a number of softwoods, such as Hemlock, traumatic (wound induced) resin canals can be seen occasionally, or even frequently. These are usually much larger than normal resin canals and show as darker lines running along the grain; they look rather like a pencil line drawn along the wood. Normal resin canals may show as faint brownish lines along the grain but are normally very hard to see.

The only other feature worth mentioning is whether the change from earlywood to latewood is gradual or abrupt. If latewood is not distinct (Parana Pine) then no change at all is visible.

Plantation-grown material is usually lighter in weight and colour; it also has wider growth rings and more knots because it is younger than trees from a natural forest. The descriptions given refer to forest-grown timber.

For many species more than one name is available. Wherever possible an entry is placed alphabetically under its standard trade name; other names are listed if they are likely to have been used in the United Kingdom.

Weights are given in metric terms and are quoted for 15 per cent moisture content. If the figure is required in Imperial terms (lb/foot^3) then it should be divided by a conversion factor of 16.

Durability comments relate to heartwood exposed to fungal decay. Sapwood is almost always perishable. Most softwoods are resistant to treatment or at best moderately resistant; this difficulty often includes the sapwood too.

Alerce
Fitzroya cupressoides

This species grows in Chile. The heartwood is a distinct brownish-red rather similar to Western Red Cedar. The grain is normally straight and it has a very fine even texture; the weight is about 510 kg/m^3 when dried.

The wood is unusual for a softwood in that it is durable; it is also easy to work and finishes well.

Alerce is used for cooperage, furniture, joinery and masts for boats. Supplies of large old trees are available which give rise to timber in large sizes and of good quality.

Cedar

(various)

Cedrus species

The common name cedar is a difficult one because it is used for both hardwoods and softwoods. When used for a softwood species it is correctly applied to various species of *Cedrus*; it is also likely to be used for other softwoods which happen to have a fragrance similar to that of the true cedars. It is often preferable not to use the word cedar alone but to qualify it in some way simply to avoid confusion.

There are three common species of *Cedrus* available: *Cedrus atlantica* (North Africa) provides Atlantic or Atlas Cedar, *Cedrus libani* or *Cedrus libanotica* (Middle East) provides Cedar of Lebanon and *Cedrus deodara* (India) provides Deodar. All the species are grown as ornamental trees in the United Kingdom. Their timbers are not different from one another. Traumatic resin canals may be present.

The heartwood is light brown with a strong and quite persistent odour. The grain is straight and the latewood clearly visible with a gradual change from earlywood. The texture is medium-fine. The timber is easy to work and finishes well. Some warping is likely during drying. The average weight is about 580 kg/m^3 when dried.

Cedar is durable and is suitable for outdoor uses in garden furniture, fences, railway sleepers and joinery. It is also used locally for house construction and furniture.

Cedar, Barbados
Juniperus barbadensis

Of local importance only, this timber is a typical juniper but has a cedar-like odour. Under drying treatments the wood behaves reasonably well and degrade figures are not too high, though there may sometimes be a tendency to warping, especially in thin stock. Knots may cause a certain amount of trouble in working the wood but on the whole it is not difficult to work and may be brought to a good surface; it needs pre-boring for certain operations. The average weight of the timber varies between 400 and 560 kg/m^3 when dried, the grain is normally straight and the texture fine and even. In colour, the wood is light red or brownish, with no distinctive figuring but with a mild and characteristic pleasant smell. Barbados Cedar may be summarized as a non-ornamental cabinet wood of local importance only.

Cedar, East African Pencil
Juniperus procera

This tree is not a true cedar but it has a marked cedar-like fragrance which is released when a pencil is sharpened.

In any but the smallest sizes, the timber of this species shows a considerable amount of surface checking, shaking and end-splitting during drying treatments but as the importance of the wood rests primarily on its use in the manufacture of pencils, the timber is normally imported into the United Kingdom in the form of pencil slats of $2\frac{1}{4}$ inches by $\frac{1}{2}$ inch end section. Pencil Cedar may be used for the lining of caskets, clothes chests and similar work where the fragrance is an asset, provided that it can be obtained in sufficiently large dimensions.

The timber is exceedingly easy to work but it splits readily and needs pre-boring for such operations as nailing and screwing. Glue adheres well and the wood will respond satisfactorily to most of the normal types of finishing treatment. Its weight is about 580 kg/m^3 when dried, the grain is straight and the texture fine and even. In colour the wood is reddish-brown and it has neither figuring nor distinctive taste. The heartwood of the species is durable and the timber is also said to be resistant to termites. The wood is used in East Africa for furniture and joinery also.

A closely related species is the Virginian Pencil Cedar (*Juniperus virginiana*). This was formerly the most important of the Pencil Cedars but supplies are now not so abundant.

Cedar, Port Orford

(Lawson's Cypress)

Chamaecyparis lawsoniana

This tree tends to be of somewhat shrubby growth in the United Kingdom but grows to timber size in the United States, chiefly in the coastal areas of California and Oregon. Alternative titles for the timber include Oregon Cedar.

It is a durable wood which dries well and has good stability standards once it has dried. It may be used for all normal softwood purposes but because of its persistent pungent smell it is specially favoured for the lining of moth-proof chests and cabinets. It is also used in boat-building and is regarded as the standard wood for use as battery separators. Its durability suits it for cladding and shingles.

The heartwood is a pale pinkish-brown colour, the grain is straight and the texture moderately fine and even. The growth ring figure is not particularly prominent. Average weight for dried stock is 500 kg/m³. Port Orford Cedar usually works well but occasional samples may be slightly gummy, leading to some difficulty in finishing processes.

The latewood in this species is narrow and inconspicuous with a gradual change from the wide earlywood.

Cedar, Western Red (Giant Arborvitae, Red Cedar)
Thuja plicata

This wood is well known for its durability and in particular for its use as split or sawn roofing shingles. The durability under the less cold and dry conditions of the United Kingdom is not quite so good and some preservative assistance is often desirable. Like Oak and a few other species this wood will stain black when in contact with iron under damp conditions; it is acidic and will eventually corrode the iron so galvanized or coated fixings and fittings should be used, or some other metal.

The sapwood is narrow and distinct. The heartwood is at first very variable, ranging from a light salmon pink to a dark brown and often showing mixed zones. When it is dry the wood becomes an even, warm mid-brown colour; after exposure to the weather for some time the brown is lost and the colour becomes silver-grey on the surface. The grain is straight, the texture rather coarse and the growth rings show quite distinctly.

Western Red Cedar is rather soft, the weight being about 390 kg/m^3 when dried; it has a distinctive smell which gradually reduces after drying. Bruising can occur easily during working and chipping of edges happens readily. Sharp tools are a necessity. Drying presents no problems except in large sizes and the wood is stable.

Apart from shingles and cladding the wood is used for greenhouses, house extensions, bee-hives and fencing.

Cedar, White

(Northern or Eastern White Cedar)

Thuja occidentalis

Although not ranked as amongst the most important timbers on the world markets, White Cedar is of considerably more than local importance. The shrinkage factors of the wood are low and in consequence it dries with little distortion. It is rated as durable and is therefore popular for outdoor use. Like all softwoods it is immune from powder-post beetle attack. On the North American continent the wood is commonly used for such purposes as roofing shingles, boat and canoe building, fencing, telegraph poles and garden furniture. The working qualities of the timber are good, both under hand and machine operations. The average weight is about 340 kg/m³ when dried, the grain is straight and the texture is fine and even. In colour the wood is a very pale brown. The latewood is narrow and very inconspicuous.

Cedar, Yellow
Chamaecyparis nootkatensis

(Alaska or Pacific Coast Cedar,
Yellow Cypress)

The average weight of this timber is about 500 kg/m^3 when dried, the grain is straight and the texture fine and even. The timber is pale yellowish in colour and has a markedly strong odour when freshly felled but this does not persist. The tree producing the wood is common in Western North America, reaching its best development in Alaska, with a geographical range as far south as Oregon. It is durable and also shows a marked resistance to the action of acids, hence it is very suitable for laboratory bench tops and similar purposes. The shrinkage factors of the wood are low and it dries with little difficulty, being very stable once the drying process has been completed. It is exceptionally easy to work in all hand or machine operations, turns and carves well, takes nails, screws, glue and finishes well. A close and similar botanical relative is Port Orford Cedar (*Chamaecyparis lawsoniana*). The stability and durability of the wood suit it for good quality exterior joinery, doors, boat-building, drawing boards, greenhouses and shingles.

Cypress, Southern (Bald or Swamp Cypress)
Taxodium distichum

The wood has several alternative names such as Yellow, White, Black and Tidewater Cypress in America but commercially two main types are recognized: Red and White Cypress. Red Cypress grows in coastal swamp areas and White Cypress grows farther inland. The darker timber is regarded as the best quality. For a softwood the timber has excellent standards of durability and is used for structural purposes, piling, flooring and a wide range of internal work; it is also used in the cooperage industries. The timber shows a considerable colour range but is usually yellowish-brown with a strongly marked growth ring figure. The grain is straight, the texture coarse and even and the average weight about 510 kg/m^3 when dried. Only rarely does the wood prove troublesome to work and then it is usually as a result of the abrupt change from earlywood to latewood. It may be given all the customary types of finishing treatment with satisfactory results. Drying is often difficult; to avoid splits and checks, if it is to be kiln dried it is advisable to give a preliminary air drying treatment. The wood comes from the south eastern regions of the United States and is used for joinery, mining timbers, sleepers and shingles in addition to the purposes listed previously.

Fir
Abies species

The name fir is commonly used for a number of species of the genus *Abies* from both Europe and America; it is often used with a geographical or other word to qualify it further.

Fir is also used for some other softwoods and this can be confusing – Scots Fir is really Scots Pine and Douglas Fir is not a fir at all.

Of the true firs the most frequently seen is the Silver Fir (*Abies alba*); it is sometimes called Whitewood, a name better reserved for *Picea abies*, the European Spruce.

Silver Fir is native to Central and Southern Europe. The timber is very similar to that of *Picea abies* but is less lustrous. It has a very faint yellowish tint and the growth rings are hard to see. The grain is straight, the texture fine and the weight about 480 kg/m^3 when dried. It is non-durable, easy to work and of reasonable strength for a softwood. It is used for general joinery, furniture, packaging and plywood.

Abies amabalis provides the Amabalis Fir from Western Canada and America; *Abies balsamea* grows in Eastern and Central Canada and America. Both of these species are lighter in weight at about 410 kg/m^3 when dried but again resemble Spruce. They may be sold individually but are often imported mixed with the slightly stronger species Western Hemlock and Canadian Spruce.

Other species, such as *Abies grandis* (Grand Fir) and *Abies procera* (Noble Fir), also from Western Canada and America, are occasionally seen. All the Firs have very similar and almost interchangeable timbers well suited for general interior joinery and carpentry, light structural work and packaging.

Fir, Douglas
Pseudotsuga menziesii

(Columbian or Oregon Pine)

This tree grows in the northern and western United States of America, British Columbia and Alberta. Douglas Fir is the most important timber of the North American continent and one of the best-known softwoods in the world. The timber comes from a tall tree and a very high percentage of the wood is free from knots. In drying there is no trouble with the timber as regards checking, twisting or cupping and the freedom from shakes is very marked. The wood is moderately durable in exposed positions but to ensure adequate penetration the wood needs to be exercised for preservative treatments by pressure processes. Douglas Fir works well in all operations provided tools are sharp. The heartwood may range from reddish-brown to a definite yellow in colour and a distinctive figuring is present in flat-sawn timbers. The latewood of each growth ring is much darker and harder than the earlywood and the change is abrupt. This leads to an uneven texture.

The average weight is about 530 kg/m³ when dried and the timber, while occasionally wavy or spiral, is mostly straight-grained.

Forest-grown Douglas Fir is very strong for its weight and is therefore used for heavy construction, dock and harbour work, railway sleepers, cooperage, joinery, flooring (when quarter cut), vats and tanks, house construction and laminated beams and arches. It is also extensively rotary peeled and used for plywood manufacture – probably more than any other timber.

Resin ducts are present.

Hemlock, Eastern (White Hemlock)
Tsuga canadensis

Eastern Hemlock is the product of a tree from the eastern North American continent. The average weight of the timber when dry is 470 kg/m³ and it is a light brown in colour with a reddish tinge and a very pronounced growth ring figuring; the grain is generally very irregular and the texture rather coarse. Under drying treatments the wood is very refractory and shows a marked tendency to warp and twist. It is an easy timber to work by hand or machine processes but under machine tools the irregular grain may cause a considerable degree of 'picking up'. The timber is not an important one from the viewpoint of world markets but finds local use for such purposes as packing cases, railway sleepers, structural work and carcassing.

The timber has several confusing alternative names, among which are Canadian Hemlock, American Hemlock, Spruce Pine and Hemlock Spruce.

This species is inferior to Western Hemlock in most respects.

Hemlock, Western   (Pacific or British Columbian Hemlock)
Tsuga heterophylla

This species is closely related to the preceding timber. It is sometimes marketed under the confusing alternative names of Grey Fir or Alaska Pine. It is a straight-grained, fine- and even-textured wood with an average weight of 500 kg/m^3 when dried and a pale brown heartwood with a fairly distinctive growth ring figuring. The wood is well known to timber consumers and is now much used as an alternative to Redwood. The heartwood of the species is classified as non-durable and is resistant to treatment with wood preservatives. The timber dries slowly but works well (though it may chip out during mortising and having a tendency to split, it should be pre-bored before nailing or screwing. Satisfactory results can be obtained with almost all the usual kinds of finishing treatment. The wood has a slight natural lustre and is in most respects much better than Eastern Hemlock.

Many pieces of this timber show occasional narrow greyish or black lines running along the grain on machined surfaces – they look rather like a line drawn with a pencil. These lines are traumatic resin ducts and will not bleed through finishes.

Juniper
Juniperus communis

This species in itself is not an important source of timber supply, though it belongs to a genus that includes several useful commercial woods, usually described as cedars. It is the timber of a small bush or tree of Europe and the northern parts of Asia and is a brownish coloured wood with a mild and fragrant smell. It is an easy working wood, though somewhat brittle but its degree of natural durability is high. Juniper creates no special problems during drying processes and does not warp to any extent. Although it is frequently somewhat knotty the wood may be used for veneering and turnery but the commercial importance of the tree rests on the Oil of Juniper it produces, used in perfumery and in the flavouring of gin. Juniper has a straight grain, a fine and even texture and weighs about 480 kg/m^3 when dried. The wood is only available in small sizes and may be used for decorative turnery, cutlery and knife handles.

Kauri, New Zealand (Kauri Pine)
Agathis australis

The heartwood of this timber is rather variable in colour – it may be pale greyish brown but when more resin is present it is a dark reddish-brown or yellowish-brown. The resin does not bleed or affect finishes. The growth rings are indistinct so the texture is particularly fine and even; the grain is usually straight and the weight about 580 kg/m³ when dry. Drying may or may not lead to warping, depending on the quality of the material but working gives no problems unless grain irregularities are present. The timber is little used outside New Zealand where the best grades are used for vats or boat-building; poorer qualities are used in general construction.

Kauri, Queensland
Agathis species

Agathis robusta may be called South Queensland Kauri; *Agathis palmerstonii* and *Agathis microstachya* are called North Queensland Kauri. All three may be called Queensland Kauri Pine. Supplies of these species are very limited so they are unlikely to be seen outside Australia. The timber weighs about 480 kg/m³ when dried so is lighter and weaker than New Zealand Kauri; it is also non-durable, easier to dry and paler in colour. Its uses include joinery, packaging, pattern making and plywood.

Larch
Larix decidua

Larch is the product of a tree of the mountainous regions of southern and central Europe and parts of Asia. The tree is of rapid growth habit and may reach to a height of over 30 metres, living for a period of more than two centuries, though it may be felled for timber when about forty years old. It is a moderately heavy timber, its range being from 480 to 640 kg/m^3 when dried, has a reasonably fine and even texture and is usually straight-grained; the wood is reddish-brown to brick red in colour, with a distinctive growth ring figuring. Under drying treatments the timber is apt to warp and shrink considerably but it is markedly free from knots; it is regarded as being amongst the most naturally durable of all the conifer timbers and is often favoured for exterior work in preference to other softwood species. In fact it is only moderately durable.

The tree is also of importance in that it yields Venice Turpentine which has a medicinal value, and also tannin. The living tree is often attacked by the fungal disease known as Larch Canker which causes early death.

Larch has been planted on a fairly extensive scale and mature timber reaching the market is of excellent quality but not as good as mountain-grown materials. A closely related species is the Japanese Larch (or Red Larch) – *Larix kaempferi*, which is a little lighter in weight but otherwise similar.

Larches are used for boat-building, fences, posts and exterior construction work. Younger stock is also used for transmission poles, pit props and piles.

Larch, Tamarack

(Tamarack, Eastern Larch)

Larix laricina

This timber is said to be the strongest of all the softwoods native to eastern Canada, as well as parts of America. Its natural durability is as good as that of the other commercial larches and the timber reacts satisfactorily to wood preservative processes, either of the tar-oil or soluble salts types. The timber has an average weight of about 580 kg/m³ when the wood is dried, the grain is normally straight and the texture moderately coarse but even. The heartwood of the species is reddish-brown in colour and the wood has a noticeable growth ring figuring.

Tamarack Larch is not important on the general world markets but it finds extensive use in Canada, especially for railway sleepers, outdoor uses and packaging. Distortion under drying treatments is slight. The working qualities of the timber are quite good but being coarser in texture than most of the commercial larch species it will not turn well; dulling of tools is not particularly marked.

Larch, Western
Larix occidentalis

Like the other *Larix* species this timber is classed as very strong for a softwood and it is also moderately durable against the attack of wood-rotting fungi; it may also be offered for sale under the name of Western Tamarack. While undergoing drying Western Larch may be refractory – checks and warping are quite common defects to be found in converted timber of the species. The working qualities of the wood are good but not exceptional. It is similar in weight to European Larch and the grain and texture are also comparable. In colour it is a reddish-brown and it has a very distinctive figuring caused by the noticeable contrast between the earlywood and the latewood.

The timber is not of any great importance for export purposes but in western United States of America and Canada it is used for railway sleepers, flooring, piling and plywood as well as structural timber.

Pine, American Pitch   (Longleaf Pine)

Pinus palustris and *Pinus elliottii*

This very resinous, highly popular timber finds considerable use in all parts of the world for such items as piling, ornamental joinery, school and church furniture, structural work and many similar uses. It is the heavier grades of *Pinus palustris* and *Pinus elliottii*, averaging about 670 kg/m^3 when dried. The lighter weight wood is marked together with other lighter species as Southern Pine.

American Pitch Pine normally has straight grain and fine and even texture. The wood is a darkish-yellow to a light reddish-brown in colour, with a prominent and distinctive figuring caused by the dark latewood in the growth rings. American Pitch Pine is not difficult to work, whether by hand or machine processes and does not unduly blunt tools but the resin content of the wood is so high that there is often considerable clogging of tools. The shrinkage factors of the wood are high but it dries with little degrade. The timber has the advantage of being readily available in long lengths. The wood is in demand on all markets but normally there is little difficulty in fulfilling orders.

The lower density grades sold as Southern Pine are lighter in weight, coarser in texture, have much wider sapwood and are significantly weaker. They are used for joinery, packaging and light construction.

Pine, Australian Brown (She Pine)

Podocarpus elatus

This species is not a true pine but is a softwood that really merits the adjective 'soft' in its physical rather than botanical meaning. It is easy to work, whether by hand or machine and its dulling effect on tools is so small as to be almost negligible; it finishes with a good clean surface and readily accepts surface finishes. Australian Brown Pine may be summarized as a general utility softwood that is suitable for all purposes for which a low-grade conifer timber is fitted; although it is considered in Australia to be one of the most important of the native softwoods it is not likely to be exported from there in any appreciable quantity. The wood may range from a pale brown to a darker brown in colour and it has no distinctive figuring. Average specimens weigh about 570 kg/m³ for fully dried stock. The grain is generally straight and the texture fine and even.

Australian Brown Pine, in common with other species of *Podocarpus*, might at first glance be taken to be a hardwood because it lacks the growth ring figure characteristic of most softwoods.

Pine, Blue

Pinus wallichiana

Like the true pines this Indian timber is easy to work and finishes to a good surface; it will also hold glue, nails and screws satisfactorily, will turn well and, if the grain is properly filled, may be satisfactorily polished. Blue Pine has low shrinkage factors, dries with only a little degrade and is reasonably stable once the drying process is complete. Like all conifer timbers it is immune to powder-post beetle damage but it is non-durable, though it may be easily treated by wood preservative processes. The wood may be infected by sap-stain fungi. The timber is used for joinery, flooring and similar softwood purposes but is weaker than Redwood and needs selection for use as a structural timber. It is a straight-grained, medium fine and even-textured wood with an average weight of 512 kg/m^3 when dried. The sapwood and the heartwood are readily distinguishable, the heartwood being a light pinkish-red with a slightly darker figuring. The wood is soft for a pine and is not much seen outside India. Resin canals are present and the transition is gradual from earlywood to latewood.

Pine, Canadian Red (Norway Pine)
Pinus resinosa

Canadian Red Pine may additionally be offered for sale on the North American markets under the titles of Ottawa Red Pine, Red Pine or Quebec Red Pine. The timber is sometimes described as the North American equivalent to Baltic Redwood and provides an exceedingly useful and commercially valuable timber. The name Red Pine comes from the red colour of the bark, as it does in Redwood. The texture is medium and even, the grain almost invariably straight and the average weight is about 450 kg/m^3 when dried. It has a broad sapwood that is pale yellow or white tinged with yellow; the heartwood is pale brown and may show a pale reddish tinge and there is a pronounced colour difference between the earlywood and the latewood. The timber has a faint but definite taste and smell of resin. Being a true member of the pine family the wood is not naturally resistant to decay but may be easily treated with wood preservatives. Canadian Red Pine is easy to work or carve and finishes with a smooth lustrous surface, while it has the additional advantage that little twisting, warping or checking of the timber occurs during drying. Resin canals are present but resin is not normally troublesome.

Pine, Caribbean Pitch (geographical prefixes)
Pinus caribaea

Pinus oocarpa is an allied species which grows on the Central American mainland; the two are often mixed together and may be sold as Caribbean Longleaf Pitch Pine.

The wood has an average weight when dried of about 710 kg/m^3, is straight grained and has a coarse, even texture. In colour the wood is yellowish to reddish-brown and has a very distinctive growth ring figure due to the prominent latewood which is very dense and occupies about half the volume of each growth ring; the change from earlywood to latewood is abrupt. The wood has a strong resinous smell and sometimes odd patches look as though soaked in resin.

This timber is slow-drying, especially in large sizes and some care is needed to avoid degrade. It is moderately durable. No particular difficulty is experienced in working the wood unless more than the usual amount of resin is present. Resin may also cause some difficulty in applying finishes.

The wood is used in boat-building, for harbour and jetty work and vat making. It can be used for general carpentry and joinery and is frequently seen in schools and churches.

Resin canals are present and are very obvious since they normally contain dark brown resin.

Pine, Cypress

(Murray Pine)

Callitris glauca

This species is not a true pine or a cypress and does not have resin canals. Cypress Pine is a naturally very durable general utility timber that is of considerable local importance wherever it may be found in Australia but although the tree exists in commercial quantities it does not reach the world markets in any appreciable quantity. The wood weighs between 550 and 700 kg/m^3 for average fully dried specimens and also has a moderately fine and even texture with a grain that is generally straight. The heartwood of the species is a lightish-yellow in colour, with a red or a brownish figuring and there is a rather distinctive smell of camphor with the timber. Cypress Pine may be found somewhat difficult to surface cleanly, owing to the number of small knots usually found in it and it needs pre-boring for nailing, screwing, mortising and similar operations as it is very apt to split, more especially when thin stock is being used.

The timber is frequently used for piling and other under-water work as it has been found to be very resistant to Teredo (ship-worm) and Limnoria (gribble) attack. The wood is also used for house construction in termite-infested areas.

Pine, Hoop (Queensland Pine)

Araucaria cunninghamii

This species is not a true pine and has no resin canals. Hoop Pine has close botanical relatives in the fairly important Chile and Norfolk Island Pines. It is light in weight, averaging about 560 kg/m^3 when dried, has a straight grain and a texture that is normally fine and even. The species ranges from pale brown to a light yellow brown in colour and may show slight pinkish or red tints; a distinctive smell, very similar to that of cheese, is present in the wood but it has no noticeable taste. Very small knot-like marks called leaf traces are often present. Hoop Pine has sometimes been described as the only really good softwood of Australia and is the product of a tree native to Queensland, New South Wales and New Guinea. It is not often to be seen on the general markets as the annual out-turn is almost entirely consumed in Australia. The timber has the average properties of the true pines in so far as regards its working, strength and durability. It is largely employed for such purposes as framings, mouldings, panelling and joinery but in its country of origin it has also achieved importance for the manufacture of plywood and veneers.

Pine, Jack (Princess Pine)
Pinus banksiana

This is a somewhat knotty timber that is native to the North American continent. It is a true pine showing resin canals. The wood dries well and does not warp or twist to any great extent. It falls into the class of timbers described as the hard pines and it is reasonably easy to work; the grain is normally straight and the texture moderately fine and even. The wood is non-durable and the timber needs incising to ensure an adequate degree of penetration by a preserving fluid. Jack Pine has an average weight when fully dry of about 500 kg/m³. In colour the heartwood may range from pale brown to reddish-brown, showing a fairly distinctive growth ring figuring. In Canada the timber is used for such purposes as railway sleepers, boxes, shuttering for concrete and similar items but the wood does not reach the general world market in any considerable quantity. It may also be used for pulp.

Pine, Lodgepole

(Contorta Pine)

Pinus contorta var. *latifolia*

This timber cannot be regarded as commercially important when it is compared with many other *Pinus* species but on the Western North American continent the wood finds considerable use for piling, mining timbers, railway sleepers, light structural timbers and for work of a similar nature. The shrinkage factors of the wood are low and it dries with little distortion and very little degrade. Lodgepole Pine is not a really hard timber to work in most processes but on the other hand it may be more difficult to bring to a good surface on a planing machine than the majority of pines owing to the prevalence of tight knots in the timber. The wood responds reasonably well to most types of decorative finishing treatment. Average dry samples of the wood should weigh in the region of 470 kg/m^3 and it should show a straight grain and a moderately fine and even texture. There is little differentiation between the heartwood and the sapwood, the wood being a dull yellowish-white in colour. Lodgepole Pine is sometimes called Black Pine or Western Jack Pine.

Pine, Maritime
Pinus pinaster

This timber has a sapwood rather wider than is customary with most true pines. It is whitish to light yellow and moderately well defined from the light red to reddish-brown heartwood, which carries a bold growth ring figure. The grain is straight and the texture rather open and uneven. For dried stock the average weight is about 530 kg/m³. As its common name suggests, the tree is most commonly to be found near coasts, often being planted in such areas to combat soil erosion.

The tree is native to various parts of Europe, particularly the Mediterranean area; it is sometimes described as the Cluster Pine. Durability standards are rather higher than those of the average pine but the timber is sometimes excessively knotty. It is favoured for rough carpentry, pit props and other mining purposes and for some building construction.

The tree can be tapped for resin; when this is done the timber will become more resinous and is likely to resemble a Pitch Pine.

Pine, Parana

(Brazilian Pine)

Araucaria angustifolia

Parana Pine, which grows in Argentina and Brazil, is one of the few important tropical softwoods but it is not a true pine. The tree is a close relative of the Chile Pine, which is well known in the United Kingdom as the Monkey Puzzle.

Parana Pine is one of the best and most commercially important timbers of its region. The wood has several local names, of which Pinho and Pinho do Brasil are perhaps the most important. The species can be used for structural work, interior joinery, flooring and most other purposes for which a softwood is normally used, although the wood does not have what might be described as the normal soft-wood appearance. The sapwood is light in colour and is usually fairly clearly marked off from the heartwood. The heartwood has a yellowish or yellowish-brown background colour with no prominent growth ring figure but is nearly always marked with distinctive bright red or brownish streaks. The grain is straight and the texture is, for a softwood, fine and even. Parana Pine has no appreciable smell and when fully dry weighs about 550 kg/m³. This is a rather difficult timber to dry and it may warp or split severely if care is not taken during the process. The wood can be dried successfully in a kiln but may still distort during machining. Strength properties are rather above average for a softwood of its weight.

The species is not resistant to fungal attack and the sapwood may develop blue-stain or be damaged by pinhole borers. It is a difficult timber to treat satisfactorily with preservative fluids. Parana Pine works easily and well in all normal wood working processes; it stains easily and can be given a good polish. The wood is not normally resinous but resin may sometimes affect machining.

The growth rings are almost invisible and resin canals are absent.

Pine, Ponderosa

(various)

Pinus ponderosa

Other common names used for this species are Western Yellow Pine, British Columbian Soft Pine and Californian White Pine.

Ponderosa Pine is a soft and resinous pine that is a native of the regions round the Rocky Mountains. Is is a non-durable timber in exposed situations and the sapwood of the species is susceptible to various types of insect and fungal staining attack. Ponderosa Pine works well under all operations and finishes with a good clean surface; its acceptance of finishing treatments is on a par with those of other commercial softwoods. The drying qualities are good and once the timber has dried out properly it is more stable than the average wood, an attribute that has made it popular for pattern making or other work where freedom from movement is essential. The wood is light in weight, dry material averaging 480 kg/m^3, the grain is straight and the texture usually fine and even; its colour is dark yellow to reddish-brown, with a distinctive growth ring figuring. There is a lot of resin in the heartwood and this may give trouble in machining; the sapwood is wide, very pale, free of resin and rather lighter in weight.

The sapwood is good for pattern making and because it resembles Yellow Pine it is also used for furniture and turnery. The heartwood is used for general construction, joinery and packaging.

Pine, Scots (Redwood)

Pinus sylvestris

The name Redwood is usually applied to material imported to the United Kingdom from European sources and a wide range of geographical prefixes such as Polish, Archangel, Baltic, etc. are likely to be used. The wood may also be known as Red Deal or even Scots Fir. Scots Pine is used for trees grown in Britain and it is one of the best known of the homegrown timbers.

The tree providing the wood may grow to a height of 30 metres or more in favourable conditions and may be felled for timber when it is between 50 and 60 years old. Timbers of good size are available as the girth of the tree is fairly large. For external work Scots Pine is rated as non-durable and only the sapwood is easily permeable. Drying presents no problems, except that if it is too slow sapstain may develop quite rapidly but in general its other qualities are average to good for its weight of 510 kg/m^3 when dry. The wood may be used for the usual softwood purposes such as joinery, construction, poles, posts, props and plywood. The timber has a reddish cast, with a distinctive figuring caused by the growth rings. Weight is variable according to the locality and nature of the soil in which the tree is grown but the grain is normally straight and the texture reasonably fine and even. As far as the United Kingdom is concerned, the best timber comes from the Highland forests but these have been exploited very thoroughly and supplies are less readily available nowadays. The timber works easily in general but knots or resin can sometimes cause problems. Although Scots Pine and Baltic Redwood both come from the same species, the timber trade generally markets them under the name indicating the origin of each – Scots Pine being normally the less highly regarded of the two, probably because it is usually plantation material which is younger, weaker and knottier.

Pine, Western White   (Idaho White Pine)
Pinus monticola

This Western North American pine is a timber that is not at all refractory under drying treatments (either in air or kiln drying) and suffers little degrade; once the drying process is finished it is more free from movement than is the average wood. It is classified as being non-durable when exposed to conditions that are favourable to termite attack or wood-rotting fungi but the wood responds fairly well to preservative treatments. Typical uses for the pale straw coloured timber include pattern making, joinery, panelling, toymaking, match splints and most other uses to which a normal softwood may be put. The timber is not particularly heavy, dry stock weighing about 450 kg/m³ and it has a very distinct growth ring figuring; the grain is normally straight and it texture fine and even.

This is not the same timber as the Canadian and American White Pine (which is sold in the United Kingdom as Yellow Pine) although the two are quite similar.

Pine, Yellow
Pinus strobus

(White Pine, Soft Pine,
Weymouth Pine)

Rather a confusing number of names is applied to this timber; many Canadian and American names include the word white but in the United Kingdom the wood is invariably called Yellow Pine.

It is a lightweight species, averaging about 420 kg/m³ when dried and has a very fine and even texture. It is almost impossible to see distinct growth rings as there is no visible latewood – the species is classed as a soft pine as opposed to Redwood (medium) or Pitch Pine (hard). The sapwood is almost white and the heartwood may be creamy white to straw or faintly pink. Not much resin is present but the resin canals typical of true pines do show a little on planed surfaces. The grain is straight and knots are unusual.

The timber dries easily and well – it has very low shrinkage factors and therefore is ideal for pattern making. Yellow Pine is susceptible to both sapstain and wood-rotting fungal attack when exposed to conditions favouring their development. The timber works well under all tool processes providing cutting edges are kept very sharp and causes only a negligible amount of dulling of the tools; it is an excellent timber for carving. It may be used for most of the normal softwood purposes, though being rather weak it is suitable only for light to medium construction and is mostly used for internal joinery, doors, musical instruments, drawing boards and match splints.

Podo (Yellowwood)

Podocarpus species

This African timber is provided mostly by *Podocarpus gracilior* but *Podocarpus milanjianus* and *Podocarpus usambarensis* are included. All these East African species provide very similar timbers and no distinction is usually made between them.

Podo is of considerable local importance and has also been imported into the United Kingdom in fair quantities. The timber may be used for all the normal softwood purposes and also for plywood manufacture.

Podo might easily be mistaken for a hardwood species, as it is of a uniform yellowish shade and does not show the growth ring figure usually associated with softwoods. It is straight grained, with a comparatively fine and even texture and has an average dry weight of about 510 kg/m³.

In most of its properties *Podocarpus* approximates to the pines and the other softwoods previously described. It is not naturally resistant to decay and may suffer a certain amount of damage from wood borers; it is one of the few softwoods that are readily treated with preservatives, either by open tank or pressure processes. The principal defect in working is a tendency to split in nailing or to chip out in mortising and similar operations. Worked edges remain sharp, it is a satisfactory species for turnery purposes and will paint and polish well, though some samples tend to stain rather patchily. The strength properties are what might be expected for a timber of average weight. Both air and kiln drying give good results but stacks need weighting to avoid distortion.

The wood is used for interior joinery and fittings but requires treatment before use in construction.

Redwood (Scots Pine)

Pinus sylvestris

This species is dealt with under the entry for Scots Pine. It is unusual in that the British Standards Institution gives it two Standard Names – therefore it requires two entries in this work. The species is the only true pine that is indigenous to the British Isles and this probably accounts for the existence of two names – one for home-grown material and one for imported.

The names also help by telling us the origin of the material; since the species is widely used for construction this is important because Redwood in general is provided by older and larger trees and the wood is therefore stronger and less knotty. Homegrown stock is provided mostly by relatively young plantations and the logs are therefore smaller and younger; it follows that the timber is somewhat weaker and is certainly knottier.

Rimu (Red Pine)
Dacrydium cupressinum

Rimu, which may also be offered for sale under the title of New Zealand Red Pine, is provided by a tree that is indigenous to New Zealand and one that may grow to a height of over 30 metres, thereby providing timbers of good dimensions. When it is freshly felled the straight-grained timber has a reddish-brown tint but with age and exposure to the atmosphere this fades to a lightish-brown and it shows a distinctive figuring made up of dark and light streaks. The timber weighs about 610 kg/m^3 when dry, is easy to work under all hand and machine processes and is rated as moderately durable. It is chiefly used for such items as cladding, panelling, flooring, furniture and interior fittings. Small quantities of the timber have been imported into the United Kingdom from time to time but it is not of any great importance to our markets, though it is one of the most important of the timbers native to New Zealand.

A slightly lighter species (*Dacrydium franklinii*) grows in Tasmania where it is known as Huon Pine.

A third species of slightly lighter weight grows in Sarawak and has been exported. It is known as Sempilor (*Dacrydium elatum*).

Sequoia   (Californian Redwood)
Sequoia sempervirens

This species grows in the west of North America and is quite distinct from the Redwood of Europe.

Sequoia may grow to 100 metres or more in height and 5 metres in diameter. The bark is noteworthy in that it is soft, fibrous and 300 mm thick or more. The wood has a narrow white sapwood and a heartwood which is a warm red-brown with an obvious growth ring figure. The grain is straight and the texture variable depending upon how well the latewood zones are developed.

The non-resinous and durable timber dries readily without degrade and is easy to work but requires care in drilling, moulding and planing owing to its softness (420 kg/m³ when dried) and tendency to splinter; really sharp tools are required.

The wood can be used for many outdoor purposes because of its durability. It has been used extensively for organ pipes. Other uses include shingles, cladding and interior trim where it will not be bruised too easily. The bark is used in certain fibreboards and for filtering processes.

The wood is carefully conserved by restricting the amount that can be felled annually. In this way the redwood groves will be retained but the timber supplies will be limited.

A related tree was once named *Sequoia gigantea* but is now more correctly called *Sequoiadendron giganteum*. It is the Wellingtonia, famous for being the largest living thing and one of the longest living, that has ever existed on earth. The timber is not useful commercially.

Spruce, Eastern Canadian   (Western White Spruce)
Picea glauca

This is the North American equivalent to the Norway Spruce of Europe and, if its use in the paper-making industry and allied trades is taken into account, must probably rank as the most important commercial timber of Canada. Also known by various geographical names, it is light in weight (averaging 450 kg/m³ when dry) with a straight grain and a fine and even texture. The heartwood and the sapwood of the species are not sharply defined, the wood being pale yellow or lightish in colour. Although the shrinkage factors are on the high side the timber dries easily and with only little distortion. The wood works easily in all operations, has little dulling effect on tools but is rated as non-durable for work in exposed positions.

Typical uses for the timber include sounding boards for musical instruments, food containers, ladders, shop fittings, oars, shuttering for concrete work, light and medium construction, agricultural implements, kitchen cabinets, canoe building, paper pulp and similar items.

In addition, the two important species *Picea mariana* and *Picea rubens* provide the Canadian Black Spruce and Red Spruce respectively. These are not normally marketed as separate timbers and in all their qualities closely resemble *Picea glauca*. Some Balsam Fir may also be mixed with the others.

Spruce, Engelmann
Picea engelmanii

(Mountain Spruce)

This timber, which closely resembles the other North American spruces in its qualities, is the product of a tree native to the Rocky Mountain regions. The tree grows to a considerable height and boards of good sizes reasonably free from blemishes are readily obtained on the American market. It is not a naturally durable wood but although its shrinkage factors are on the high side it dries easily and well. Typical uses for Engelmann Spruce include the making of food containers, planking, sounding boards for musical instruments and light structural work. It is rather light in weight, average dry timber weighing about 450 kg/m³; the grain is normally straight and the texture fine and even. It is whitish or lightish yellow in colour and is without any characteristic taste or smell. The timber is sometimes marketed with other species as Spruce.

Spruce, Himalayan
Picea morinda

This timber has a weight of about 465 kg/m^3 when dried, a straight grain, a fine and even texture and the heartwood is brownish in colour. The timber ranks as one of the most important of the softwoods native to India, where it is used for such purposes as planking, water troughs, packing cases and other uses for which a cheap-grade softwood may be employed. It is normally an easy timber to work under all processes but some specimens are found to be rather knotty and may cause a certain amount of trouble in machine planing. Himalayan Spruce is non-durable for external use, not resistant to termite attack and is not easy to treat with preserving agents; like all softwoods it is immune to powder-post beetle infestation. To obtain a better degree of penetration of preservative fluids with this timber it is often incised by being passed through studded rollers. It is a close botanical relative of the other true spruces and tests of its mechanical properties made in India have shown it to be very similar in all respects to European Spruce.

Spruce, Sitka   (Silver and Tideland Spruce)
Picea sitchensis

This species grows in the Pacific coast belt of Canada and the United States of America. With a dried weight averaging about 450 kg/m^3 Sitka Spruce is slightly heavier than Western White Spruce but in all other respects it conforms closely to that wood. The timber is mostly straight-grained, fine and even in texture and in colour is a creamy white with a pale pinkish tinge, the heartwood and the sapwood of the species not being well distinguished. Sitka Spruce is non-durable and is somewhat subject to sapstain fungal infection. The timber dries well (both air and kiln) but care is needed to avoid degrade. It works very well under tools, finishing to a smooth surface and responds satisfactorily to the normal decorative finishing treatments for a timber of its class. Sitka Spruce may be used for all the normal softwood purposes like construction and joinery. It is also used for gliders, sailplanes, oars and racing sculls because of its high strength to weight ratio.

The species is widely grown on plantations in the United Kingdom. The best grades of the timber are suitable for constructional purposes but lower grades are only suitable for less demanding purposes.

Whitewood
Picea abies

(Norway Spruce, White Deal)

This timber must not be confused with the American Whitewood which is a hardwood.

The softwood given the British Standards Institution name of Whitewood is imported to the United Kingdom from various European countries and is often sold as Finnish, Baltic, Russian, etc. Whitewood. The name Norway Spruce is mostly applied to homegrown material. Imports from the more central and southern parts of Europe often include a proportion of Silver Fir (*Abies alba*).

The sapwood and the heartwood of the timber are not clearly marked off from each other, the wood being a very pale yellowish-brown colour, sometimes almost whitish. Whitewood has only a slight growth ring figure and a fine texture. It is a lightweight species, averaging when fully dry only 490 kg/m³. The planed surfaces of the wood have a slight sheen and are only very slightly resinous. The smell of dried timber is not distinctive.

This can be described as a typical softwood timber. It works well with standard equipment and worked edges remain sharp but it is not popular for such purposes as turnery or steam bending treatments. It is non-durable but it responds well to all preservative agents including the water-soluble as well as the coal tar derivatives. Drying creates no special problems except when knotty and the wood holds its shape well when dry. It may be used for all the normal softwood purposes, though it is best suited for work under cover or when protected from the soil. Generally it is the equivalent of Redwood in most respects but varies a good deal according to its origin. The very best mountain grown material is used for violin manufacture under the misleading name Romanian Pine. At the other extreme plantation thinnings are sold as Christmas trees.

Resin canals are present but not obvious.

Yew
Taxus baccata

Species of yew are indigenous to Europe, parts of Asia, the Himalayas and Burma, producing a wood that is well known on the timber markets. The wood is not readily obtainable in large sizes and may split if dried too rapidly. Like all the softwoods Yew is immune to powder-post beetle attack but unlike the majority of coniferous timbers is rated as durable when used in exposed positions. The working qualities are variable, depending upon the grain direction, whether under hand or machine operations and the timber gives good results with most of the normal kinds of finishing agent. Yew has a grain that is either straight, shallowly interlocked or just irregular, a fine and even texture and an average weight of about 670 kg/m^3 when dried. The heartwood is a light red to a deep brown in colour, some specimens showing an indistinct figuring of darker streaks. Depending upon the log there may be many small knots or bark pockets.

The wood may be sliced to produce decorative veneers using pin knots, burrs or even sapwood and heartwood mixtures as added attraction. The wood assumes a beautiful golden brown colour with age and exposure to light and is used for high quality furniture, cabinets and doors. It also turns well and is popular for archery bows.

Bibliography

A great many books and papers have been published on regional or world timbers. A selection of them is given here to enable readers to seek extra information or species if they need it. A few titles dealing with general timber technology, or particular disciplines such as drying or preservation, are included. Most of these titles can be obtained through public libraries or from the research and development organizations that produced them.

Anderson, R.H. *The Trees of New South Wales.* Government Printers, Melbourne, 4th edition, 1968.

Anon. *The International Book of Wood.* Mitchell Beazley Publishers Ltd, London, 1976.

Baker, R.T. *Hardwoods of Australia and their Economics.* Government of New South Wales, Sydney, Technical Education Series No. 23, 1919.

Baker, R.T. and Smith, H.T. *A Research on the Pines of Australia.* Government of New South Wales, Sydney, Technical Educational Series No. 16, 1910.

Boas, I.H. *Commercial Timbers of Australia.* Government Printer, Melbourne, 1947.

Bolza, E. *Properties and Uses of 175 Timber Species from Papua New Guinea and West Irian, Australia.* CSIRO, Australia, 1975.

Bolza, E. and Keating, W.G. *African Timbers: Properties, Uses and Characteristics of 700 Species.* Melbourne Division of Building Research, CSIRO, 1972.

Bootle, K.R. *Wood in Australia: Types, Properties and Uses.* McGraw-Hill, Sydney, 1980.

British Standards Institution. *Nomenclature of Commercial Timbers, Including Sources of Supply.* BS 881 & 589, London, 1974.

Brown, H.P. and Panshin, A.J. *Commercial Timbers of the United States.* McGraw-Hill, New York and London, 1940.

Brown, H.P., Panshin, A.J. and Forsaith, C.C. *Textbook of Wood Technology.* McGraw-Hill, New York, 1949.

Bryce, J.M. *The Commercial Timbers of Tanzania.* Moshi, Forest Division, 1967.

Building Research Establishment (Princes Risborough Laboratory)
Handbook of Hardwoods, 3rd edition, HMSO, 1981.
Handbook of Softwoods 3rd edition, HMSO, 1983.
Timber Drying Manual. BRE Report, HMSO, revised 1986.
Identification of Hardwoods. A Lens Key, Forest Products Research Laboratory, Bul. No. 25, HMSO, London 1960.
An Atlas of End-Grain Photomicrographs for the Identification of Hardwoods, Forest Products Research Laboratory, Bul. No. 26, HMSO, London 1953.
The Strength Properties of Timber, Forest Products Research Laboratory, Bul. No. 50, HMSO, London, 2nd Edition, 1969.
The Natural Durability Classification of Timber, Princes Risborough Laboratory, Technical Note No. 40, 1973.
The Moisture Content of Timber in Use, Princes Risborough Laboratory Technical Note No. 46, 1970.
Selecting Ash by Inspection, Princes Risborough Laboratory, Technical Note No. 54, 1972.
Reaction Wood, Princes Risborough Laboratory, Technical Note No. 57, 1972.

Coggins, C.R. *Decay of Timber in Buildings: Dry Rot, Wet Rot and Other Fungi.* East Grinstead, Rentokil Ltd. 1980.

Corkhill, Thomas. *A Glossary of Wood.* Stobart & Son Ltd., London 1979.

Dallimore, W. and Jackson, A.B. *A Handbook of the Coniferae.* Edward Arnold & Co., London 1948.

Desch, H.E. *Timber; its Structure, Properties and Utilisation.* 6th edition revised by J.M. Dinwoodie, Macmillan and Co., London.

Edlin, Herbert L. *What Wood is That? A Manual of Wood Identification with 40 Actual Wood Samples.* Stobart & Son Ltd, London, 1977.

Entrican, A.R. and others. *The Physical and Mechanical Properties of the Principal Indigenous Woods of New Zealand*, New Zealand Forest Service, Wellington, NZ 1951.

Findlay, W.P.K. *Timber: Properties and Uses.* London, Crosby Lockwood Staples, 1975.

Hickin, N. *The Insect Factor in Wood Decay: An Account of Wood-Boring Insects With Particular Reference to Timber Indoors.* London, Associated Business Programmes, 3rd edition, 1975.

Hora, B. *The Oxford Encyclopaedia of Trees of the World.* Oxford University Press, Oxford, New York, 1981.

Howard, Alexander L. *Timber of the World, their Characteristics and Uses,* Macmillan & Co. Ltd., London, 1951.

Jane, F.W. *The Structure of Wood,* revised by K. Wilson and D.J.B. White. Adam & Charles Black, 2nd edition, London, 1970.

Johnson, H. *The International Book of Trees.* London, Mitchell Beazley Publishers Ltd 1976.

Kribs, David A. *Commercial and Foreign Woods on the American Market.* Dover Publications Inc., New York, 1968.

Lee, Yew Hon and Chu, Yue Pen, *Commercial Timbers of Peninsular Malaysia.* Dept. of Forestry, Malaysian Timber Industry Board, Kuala Lumpur, 1975.

Lincoln, W.A. *Complete Manual of Wood Veneering* (includes a list of 250 veneering timbers), Stobart & Son Ltd., London 1984 and Chas. Scribner's Sons Inc., New York, 1985.

Lincoln, W.A. *World Woods in Colour.* Stobart & Son Ltd, London 1986.

McElhanney and Associates. *Canadian Woods: their Properties and Uses.* Forest Products Laboratories of Ottawa, 1935.

Menon, P.K.B. *Structure and Identification of Malayan Woods.* Forest Research Institute, Record No. 25, Kuala Lumpur, 1967.

Oliver, A.C. *Woodworm, Dry Rot and Rising Damp.* Sovereign Chemical Industries Ltd, Barrow-in-Furness, 1984.

Organisation for European Economic Co-operation. *African Tropical Timber: Nomenclature and Description.* OECD, Paris, 1951.

Palutan, E. *Timber Monographs.* 3 volumes each containing 40 samples of real wood veneer. Milan, Meta Publishers 1982 (English edition).

Pearson, R.S. and Brown, H.P. *Commercial Timbers of India.* 2 volumes. Govt. of India Central Publications Branch, 1932.

Pleydell, G.J. *Timbers of the British Solomon Islands.* United Africa Co. (Timber) Ltd, for Levers Pacific Timbers Ltd, London 1970.

Record, S.J. and Hess, R.W. *Timbers of the New World.* New Haven, Yale University Press, and London, Oxford University Press, 1943.

Rendle, B.J. *World Timbers.* Volume 1. Europe and Africa; Volume 2. North and South American, Central America and West Indies;

Volume 3. Asia, Australia and New Zealand. Ernest Benn Ltd., London, 1969–1970.

Richardson, B.A. *Wood Preservation.* The Construction Press, 1978.

Swain, B.H.F. *The Timbers and Forest Products of Queensland.* Queensland Forest Service, Brisbane, 1928.

Timber Research and Development Association
Timbers of Africa. Red Booklet No. 1. TRADA, 1978.
Timbers of South America. Red Booklet No. 2. TRADA, 1978.
Timbers of Southern Asia. Red Booklet No. 3. TRADA, 1978.
Timbers of South East Asia. Red Booklet No. 4. TRADA, 1978.
Timbers of Philippines and Japan. Red Booklet No. 5. TRADA, 1978.
Timbers of Europe. Red Booklet No. 6. TRADA 1978.
Timbers of North America. Red Booklet No. 7. TRADA 1978.
Timbers of Australasia. Red Booklet No. 7. TRADA, 1978.
Timbers of Central America and the Caribbean. Red Booklet No. 9. TRADA, 1978.

Trotter, H. *The Common Commercial Timbers of India and their Uses.* (Revised edition) Govt. of India Press, Calcutta, 1941.

U.S. Dept of Agriculture, Washington, D.C. *Present and Commercial Timbers of the Caribbean*, Agriculture Handbook No. 207, 1971.

United States Forest Service, *Wood Handbook.* Forest Products Laboratory, US Dept. Agriculture No. 72, Washington, 1955.

Webster, C. *Timber Selection by Properties: the Species for the Job. Part 1: Windows, Doors, Cladding and Flooring*, HMSO, 1978.

Webster, C., Taylor, V. and Brazier, J.D. *Timber Selection by Properties: the Species for the Job. Part 2: Furniture*, HMSO, 1984.

Wimbush, S.H. *Catalogue of Kenya Timbers*, Nairobi Govt. Printer, 1950.

Index